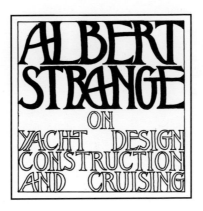

ALBERT STRANGE ON YACHT DESIGN CONSTRUCTION AND CRUISING

Jamie Clay and Mark Miller

Preface by Maurice Griffiths GM

THE ALBERT STRANGE ASSOCIATION

To the memory of Herbert H Reiach, MINA
1883–1921
Founding Editor of *The Yachting Monthly*

His magazine created a forum of inestimable value
to the development of small boat sailing.
Sadly his untimely death prevented him
from carrying out his intention to publish a book
about the designs and writings of his friend,
Albert Strange.

First published in 1999 by
The Albert Strange Association
Saxon House
83 Ipswich Road
Woodbridge
Suffolk IP12 4BT
UK

ISBN 0 9526160 0 9

Designed and produced by
Krista Taylor, KT Design, Falmouth, Cornwall.

Repro by Scantec, Falmouth, Cornwall.

Printed and bound in the UK by
TJ International Ltd, Padstow, Cornwall.

Copy edited by Jenny Bennett.
Proof and index by Rob Maynard.

Maps and title typography by Tony Watts.
Cover design by Krista Taylor.

*Cover: Painting from the Albert Strange Association archives,
attributed to Albert Strange.*

PREFACE

In yachting circles Albert Strange's name has long been associated with the development of beautiful designs of small cruising yachts. His exquisitely drawn plans show the hand of an accomplished artist and it is thought that he could have successfully made naval architecture his profession.

He had been trained from an early age as an artist however, and in this profession he became Headmaster of the Scarborough School of Art when it opened in 1882. At the same time he pursued his boat designing as a passionate hobby.

Strange took a keen interest in the sailing and racing of model yachts and became a founder member and the first Vice-Commodore of the Scarborough Model Yacht Club in 1887, where he soon showed a distinct ability to design and build successful racers – some with bermudan mainsails – whilst his new ideas brought many developments to the sport. At the Scarborough School of Art, Albert Strange was an able and popular teacher and it was said that if any of his pupils showed an interest in boats and sailing he would give them lessons in design and the drafting of plans.

From the pointed stern of his sailing canoes, with the rudder hung on the sternpost, Strange developed what became known as a canoe stern with the rudder set inboard. In a number of these designs his artistic sense produced variations in canoe sterns, from well-rounded in profile and chubby in section, to the long drawn-out profile like a counter brought to a point. With all these variations he managed to match a shapely bow with a pleasing sheer which brought to yacht design a new concept of elegance, so that an Albert Strange yacht could be picked out in any anchorage.

For the benefit of a larger audience Strange compiled a series of articles explaining the niceties of calculations involved in designing a yacht, and these were published in *The Yachting Monthly* throughout 1914 and 1915. The complete series is reproduced in this volume. Although compiled some 80 years ago, the methods involved still hold good for cruising yachts today. In our present electronic age the work of the professional naval architect has developed into a high-tech sphere where a vessel's lines plans are no longer drafted by hand with guidance from ship's curves, lengthy splines, numerous lead spline weights to hold the curves steady, planimeter, and other traditional instruments of the marine architect's art, but are now flashed onto a screen where they can be instantly modified at will and the calculations checked in seconds. Nevertheless, the fundamental principles of naval architecture, the calculations and balance of weights, the study of buoyancy and stability and displacement of a vessel are still bound by the same laws for a little yacht as they are for a big freighter. Strange's articles are as sound in principle today as they were when he wrote them and the reproduction of them here is to be applauded.

Of various cruising exploits, first in sailing canoes and later in cruising canoe yachts, Strange wrote a number of enjoyable accounts which became a feature in *The Yachting Monthly*'s early years. Illustrated with his attractive and often stirring pictures they were a delight to many readers; and here in this volume are reprinted three of the best yarns.

Today's complicated plastic, steel or light-alloy vessel, lavishly equipped with the latest gadgets in rigging, navigation, telecommunications and comfort, offers the modern yachtsman advantages undreamt of in Strange's day. Yet, in reading these pleasing and often whimsical yarns of Strange's boating forays, today's yachtsman will surely be captivated by the simplicity of cruising in very small craft without motors, and the courage with which their skippers would face emergencies, with no thought of calling out a lifeboat.

The spirit of sailing on one's own wild lone, or with a good shipmate, lives on in these pages for the delight of all those who today go afloat in little ships. May their voyages be blessed.

Maurice Griffiths (1902-1997)
West Mersea, Essex

ACKNOWLEDGEMENTS

Many people have helped in various ways with the task of putting this book together. Jamie is indebted to Kevin Fuller of Maldon and John Leather of Fingringhoe who kindly made their libraries available to him; and to his brother Pete Clay, and Jane Buckley who offered useful comments on the manuscript.

Two yacht designers and fellow boatbuilders, Arthur Holt of Maldon and Fabian Bush of Rowhedge, gave him valuable assistance with the critique of the four yawls in Chapter One.

Thanks are due to Tony Holdich of Cherry Burton for providing information on the early days of the Humber Yawl Club, and to Roderick Kalberer of Osea Island for details of various one-design classes. Hilary Peyton lent a professional hand with the drawing on page 14.

Jamie acknowledges a deep debt of gratitude to his late father, Jim, whose great patience and generosity engendered, from a very early age, his love of boats, and who also allowed him, as an inexperienced teenager, to skipper his beloved yawl, *Firefly*.

Mark would like to thank Mystic Seaport Museum, USA, for the drawings on pages 130, 132 and 163 and Mike Burn, owner of the Albert Strange designed *Sheila I*, for the loan of rare early numbers of the Humber Yawl Club Yearbooks.

He also wishes to record his appreciation to his wife, Priscilla, for her advice and support throughout the project.

We are grateful to the late Maurice Griffiths, editor of *The Yachting Monthly,* 1926–1967 and a designer of many successful cruising yachts, for contributing the preface.

Bill James, long-serving former Hon. Secretary of the Albert Strange Association, put much work into the early days of the project, while Tony Watts has kindly provided maps and other most valuable help. Jenny Bennett's copy editing and Krista Taylor's design and co-ordination of production of this book have refined and expedited it to publication, and we thank them.

CONTENTS

★ *To assist the reader, some sub-headings have been added to the original text, and are enclosed in brackets.*

FOREWORD

In 1990 the Albert Strange Association commissioned the publication of a book, *Albert Strange, Yacht Designer and Artist, 1855-1917* ★. The author was John Leather, a writer noted for his many works on traditional sailing craft. Much of the content was based on research undertaken by members of the Association. This help was kindly acknowledged by the author in his introduction.

The book covered many aspects of Strange's life and also placed in perspective his contribution to yacht design by reference to other successful designers of the period. Sheila Willis and Tony Watts contributed a chapter on Strange as an Artist and Tony also produced a most detailed account of Strange's involvement in the world of Model Yachting.

The book was a great success and all copies were sold — much to the relief of some members of the Association who had financed the publication. A second edition was considered but postponed in favour of this present volume which is intended as a companion to the original work. Where John Leather covered many facets of Strange's career, we have concentrated on Strange as a yacht designer. The series of articles on design which Strange wrote for *The Yachting Monthly* are reprinted in full and are accompanied by a variety of his designs, most of which were not considered in any great detail in the first book. We have also included two cruise accounts and an historical essay to reveal Strange's ability as a fine story teller.

The idea for this companion volume was first suggested in 1992. Early on we each spent some time as sole editor but little was achieved until we joined forces. Mark has done most of the research, Jamie most of the writing. But on one matter we are indivisible: we jointly accept responsibility for any errors.

Jamie Clay
Maldon, Essex

Mark Miller
Feock, Cornwall

★ *Albert Strange, Yacht Designer and Artist, 1855-1917*
Author – John Leather. ISBN 0 946270 73 2.
Publisher – The Pentland Press Ltd., Edinburgh.

Throughout this book the abbreviation 'JL' refers to the above publication by John Leather.

The *Yachting Monthly* articles which follow have been reproduced from the original printed pages and we have endeavoured to achieve the best possible quality.

CHAPTER 1

An Introduction

"SHEILA" PASSING
PLADDA LIGHTHOUSE
JULY 22. 09

ROBERT. E. GROVES. 1910.

YM Aug 1910

By the time of his death in 1917, aged 61, Albert Strange is thought to have designed some 150 craft; among extant drawings, his own numbering of the designs reaches No. 154, but is not entirely reliable. They range from small dinghies and centreboard boats to yachts of 45 tons and more. The great majority, including those for which he is best remembered, lie in the 20ft to 40ft range. Thirty-five of his designs are listed in Lloyd's Register of Yachts at the outbreak of the First World War in 1914, but as many as 60 may have been afloat at that time.

The first design of which we have a record is that of *Cherub*, a 21ft transom-sterned cutter which he designed for his own use in 1888, at the age of 33. The following year, the lines of a 15ft 6in clinker boat, *Wren*, appeared in *The Field, The Country Gentleman's Newspaper*, accompanied by a detailed letter from Strange. In another weekly newspaper, *The Yachtsman*, for 14th January, 1892, appear the lines of a 'Single-Handed Cruiser' (see JL page 97), inscribed with the design number 4. These drawings are dated December 1891 and accompany the transcript of a lecture which Strange gave to the members of the Royal Yorkshire Yacht Club. These three designs appear to represent Strange's beginnings as a designer and they must hold considerable interest for us if we are to understand how his career as a designer developed.

In the RYYC lecture, Strange refers to his own sailing experience, 'I have now completed 18 summers of more or less single-handed work'. This takes us back at least to 1874. He in fact acquired his first boat in 1872, at the age of 17. This was *Dauntless* (possibly renamed from *Stella*), a double-ended clinker boat some 40 years old, of a local inshore fishing type known as a peter boat. She was about 20ft. long and rigged with spritsails on main and mizzen masts (see JL page 15). In *Dauntless*, Strange no doubt got himself into and extricated himself from many a 'scrape', learning thoroughly the ways of a small boat in the process.

His cruising ground was the lower Thames, where he was born and lived in the riverside town of Gravesend, some 20 miles below London and 15 miles above the river's mouth, where it widens into the great expanse of sands and channels known as the Thames Estuary.

In his teenage years Strange spent much time in the company of local fishermen, absorbing the busy life of the waterfront and accompanying the skipper of the bawley *Eliza* about the local fishing grounds. Occasionally he would be taken along as a boy when this particular fisherman friend shipped aboard one or other of the great racing cutters as local pilot for the spring matches on the Thames.

Strange's father tried to steer him towards a career in law, but first music and then painting gained the ascendancy. The details of this period of Strange's life are sketchy, but it is evident that he was achieving some success as an artist and is known to have spent two years as Second Master teaching at the Liverpool College of Art. In 1881, a sketching cruise was undertaken in the company of a writer friend with whom he had jointly purchased *Quest*, a 26ft ex-government cutter. This cruise, which was of eight months' duration, took them about the Thames Estuary, the south coast of England, and to France. Their return, in March 1882, forms the subject of *A Winter's Tale*, reproduced on page 221.

There is no doubt that Strange's experiences in both *Dauntless* and *Quest* had a formative influence on his own small boat designs, though it was to be several more years before he started to design. *Quest* was disposed of soon after their return. In August of the same year Strange was married to Julia Woolard. In the following month he took up his appointment as Headmaster of the new School of Art at Scarborough on the Yorkshire coast, a post in which he continued until 1916. Little is known of his sailing in the seven years up to 1888. His work load must have been heavy, while at home there were soon two children to occupy him.

Cherub, launched in 1888, was based at Scarborough. Strange states in a letter to *The Field*, in February 1891, that he had 'cruised in her for hundreds of miles up and down the East Coast, between the Firth of Forth and this place'. Later in that year, he cruised in her right down to Harwich in Essex, a distance of some 220 miles. She was unusual in that she was fitted with two centreplates, fore and aft. The greatest advantage of this arrangement was that, by judicious adjustment of the boards, 'one can leave the tiller for 10 minutes at a stretch when alone, with the certainty that the boat will keep her course meanwhile'.

Provided by Strange to accompany an account of his long summer cruise, in
CHERUB, *to Harwich and Woodbridge (where she was laid up for the
winter). Taken from the Humber Yawl Club Yearbook, 1891.*

Wren, 1889, may well have been Strange's first commissioned design. His enthusiastic letter in *The Field* which accompanied the plans, makes it clear how the design came about:

> *The main objects sought to be compassed in the* Wren *were these: (1) A fairly fast and safe little ship. (2) One that could easily be put on a steamer's deck or on a railway truck. (3) More comfortable and spacious sleeping accommodation than a canoe. (4) To be convertible into an almost open boat for day sailing with friends and to be well within the compass of a single hand, whether under canvas or when rowed occasionally in a calm or about a harbour.*
>
> *This is a fair number of requirements for a boat to fulfil, and I think the* Wren *should fulfil them all very well indeed. Before putting pencil to paper, or before even digging out the drafting board, I turned over the pages of the boat sailor's guide,* Dixon Kemp's Yacht and Boat Sailing, *to see if we couldn't take advantage of someone else's experience. But nobody seemed to have wanted quite what we desired and so we took the design of a 15ft open sailing and rowing boat therein drawn, and on that we proceeded to base our calculations. We altered the lines a wee bit, giving a wider transom, pinched her beam a bit to save what weight we could, did away with the centreboard case above the floor of our future stateroom, half-decked her, and added thereto a good 6in coaming, and then set about the crux of the whole business – to obtain a dry cabin, which should at some point give 3ft headroom when we were below, and not be too much in the way of working the boat under sail. This could inevitably only be done by something movable...*

Despite *Wren* having been based on an existing and proven design, this passage shows a good deal of confidence on the part of the designer, notwithstanding the fact that it was written with the benefit of a thorough and successful test of the new boat's capabilities – on the day she was launched at Scarborough, Strange took her out alone 'in a smart northerly breeze and a good deal of swell'. A month later, he had her shipped by coasting steamer down to Chatham, where, he relates:

> *... the owner came on board and we had a very jolly fortnight's cruise on the Medway and the lower Thames, finishing up with a trip to the Roach and Crouch. We had plenty of wind and rain the whole time, but in the August bank-holiday breeze, we found out what a really comfortable and staunch little ship she was.*

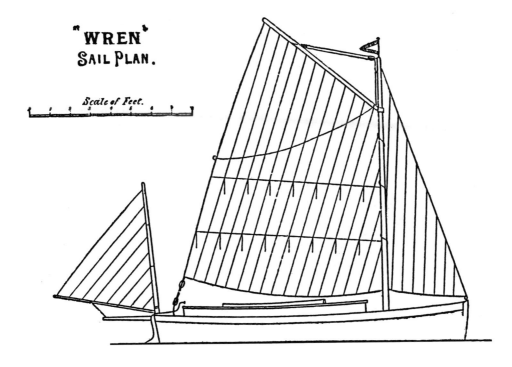

"WREN"
SAIL PLAN.

Scale of Feet.

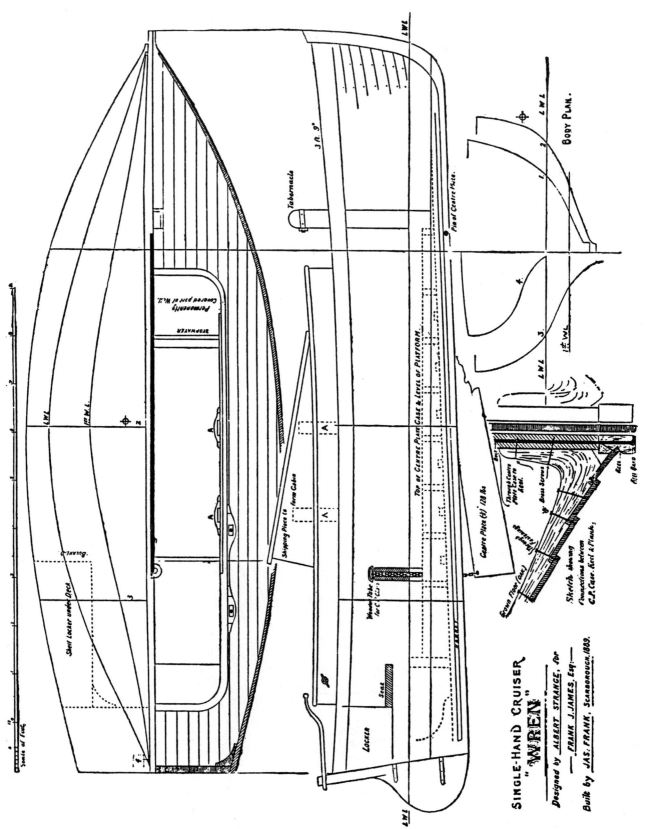

SINGLE-HAND CRUISER
"WREN"

Designed by ALBERT STRANGE, for
FRANK J. JAMES, Esq.

Built by JAS. FRANK, Scarborough, 1889.

Although the required level of comfort and security may have been achieved, the long, low sail plan and straight keel do not indicate handiness. It was not many years before Strange had left this somewhat primitive design a long way behind.

The *Wren* and *Cherub* designs were thus two, possibly the first two, designs essayed by Strange and were both thoroughly tested by first-hand experience. In addition to this grounding, Strange was by now actively engaged in the designing, building and racing of model yachts with the Scarborough Model Yacht Club, of which he was a founder member in 1887.

Design Number 4, 1891, was drawn shortly before Strange joined the Humber Yawl Club and is of a type and size then becoming popular in that club. The drawing includes alternative scales, giving a boat of either 16ft or 18ft waterline. The design (see JL page 97) has not the elegance of later work, but shows many elements which Strange was soon to refine. In particular it is the first example of what was later to be called the canoe stern; that is to say, an overhanging pointed stern with the rudder hung beneath in the manner of a counter stern. This was an innovation (although there were precedents) which was soon brought to a level of refinement by Strange which, in turn, allowed its novelty to pass into popular acceptance.

Much research remains to be done on other early designs by Strange. We have records of less than a dozen between No. 4 in 1891 and *Sheila*, designated No. 70 in 1903. *Sheila* (see page 10) is the first 'keel canoe-yacht' of which we have any knowledge. Other designs from the period included in this volume are *Mona*, *Tavie II* and *Wenda*. *Birdie* (No. 39, later *Dorcis*), designed for a member of the Humber Yawl Club in 1898, is the first keel boat. She was a conventional cutter, 29ft. overall, with a counter stern (see JL page 49).

Strange appears steadily to have widened his experience of different cruising grounds and the diversity of his output reflects this. The demand for his designs grew among friends both in these localities and through correspondence and interchange of ideas in the yachting press of the time, on a national and international level.

One such friend was the well known American yachting writer, W P Stephens. A perspective on the contribution Strange made to yacht design and cruising is given in the obituary Stephens wrote of Strange for *The Rudder* in 1917.

> *His work as a designer appealed to me as based upon actual experience at sea and a broad and comprehensive view of the whole subject of design; while recognising the value of racing as an essential part of yachting and giving much time to the study of measurement, he had no use for the racing machine, the freak or the rule-cheater.*
>
> *. . . The cruisers designed by Mr Strange, some 150 in all, were notable for their refinement of form, giving speed, handiness and weatherliness, for the amount of internal accommodation and its excellent arrangements; while his construction, though thorough in every way, lacked the excessive weight of the small conventional cruiser, with a proportionate gain in speed . . . He held a wholesome belief in the refinement of form and reasonable speed as elements of safety in a cruising yacht.*

Geo. Holmes   YM April 1909

One morning in the Autumn of 1910, Herbert Reiach, founding editor of *The Yachting Monthly*, called in at the small Scottish harbour town of Tarbert, Loch Fyne. His great pleasure at what greeted him there is evident from his report of the visit.

It was a lucky morning, for with one stone, so to speak, I took Quest II, *her owner, her designer, and her builder. Tarbert Harbour is always a pleasant place, but to drop in 'on spec.' and find men and boats of so much interest was good fortune. There lay* Cherub III, *fresh from her builder's hands ...,* Quest II *at her moorings and with her owner aboard,* Hawk Moth *was beating in single-handed from the Western Highlands, and, on the foreshore, Mr Albert Strange, designer of the three, hands in pockets, and in his element.*

The builder was Archibald Dickie, who had come to Tarbert from Fairlie, some 14 years previously. His yard no doubt serviced the large fleet of local herring boats, but the yacht building side of the business was expanding rapidly.

Quest II (see page 125f.) was a 21-ton ketch, built four years previously for C W Adderton, an artist friend and former pupil of Strange. *Hawk Moth* was a 32ft yawl, two years old, owned by the naturalist Walter Beaumont (see also page 170), and the newly launched *Cherub III* was Strange's own boat. Another artist, Robert Groves, also based his yawl, *Sheila*, here and plans were already in hand for the larger *Sheila II* (page 109) to be built by Dickie.

These men were all members of what Beaumont's obituarist later referred to as 'the little band of yachtsmen who make Tarbert their headquarters, and who join to their love of the sea and devotion to cruising either an artistic or scientific interest'. Tarbert lay at the heart of a challenging and inspiring cruising area. It attracted these men who were all experienced and competent amateur yachtsmen, with the means and the time to indulge their various interests afloat. Among their number was Strange – a designer who understood from first-hand experience the type of craft they required; and on hand was a well-respected builder. These happy circumstances provided the catalyst for some of Strange's best known and most successful designs.

Part of Dickie's Yard   *Robert Groves*   *YM Sept 1913*

Sheila   *YM July 1909*

At this time Strange was at the height of his output as a designer of small cruising yachts. The three which Reiach happened upon on that autumn day represent something of their diversity: *Quest II*, large and comfortable, sufficiently commodious for the owner's guests and their golf clubs, but with a modest sail area for handling by the owner and one paid hand; *Cherub III*, designed to a budget but with accommodation cleverly contrived for a family; and the much leaner, lighter, *Hawk Moth* for the single-hander.

The designing of a yacht involves a series of compromises. Speed, seaworthiness, seakindliness, handling, draught, good looks, accommodation, cost and other elements all have to be taken into consideration, according to the owner's priorities. The success of a particular design lies in the degree to which the correct balance between these elements is achieved. There is a great deal of wisdom and insight on the subject of achieving this balance to be found in Strange's articles on yacht design which follow this chapter.

It might be illuminating here, however, to take four of his designs, chosen for their outward similarity in size and type, and compare them with a view to seeing in which aspects they differ and what may lie behind these differences. It is as well to remember, in this age of mass production, that the designs come from an era when there was not a huge pool of existing designs, nor indeed boats, to choose from; and in any case, unless his interest was in one-design racing, or he had some very specific reason for so doing, a man would no more think of building a copy than he would eat the same meal every evening.

The four designs span 10 years between 1904 and 1913. It is tempting to study them for a development in Strange's work, but it has to be borne in mind that each design is a response to a particular owner's brief; any development is likely to be found in the way in which that brief is carried out.

The principal dimensions of the four are tabulated below and the lines plan of each has been reduced to approximately the same scale – ¼in to 1ft – to enable easier comparison. A simple method of comparing individual lines is to trace them onto tracing paper (including a datum such as the waterline or centreline), which can then be superimposed onto another lines plan. But it must be borne in mind that different plans are divided into different numbers of sections, so that the sections of the body plan, the buttock lines and the waterlines on one design will not always correspond with those of another.

	SHEILA	MIST	NORMA	THERESA II
Year	1904	1906	1910	1913
Sail Area (sq.ft)	330	409 + tops'l 63	365	359 + tops'l 50
LOA	25ft	26ft 6in	25ft 5in	25ft 6in
LWL	19ft 6in	20ft	20ft	20ft 6in
Beam	6ft 9in	7ft 1in	7ft 2in	7ft 6in
Draught	3ft 5in	4ft	3ft 5in	3ft 7½in
Displacement (tons)	2.25	3.32	3.35	3.5, approx.

An exact displacement is not available for *Theresa II*

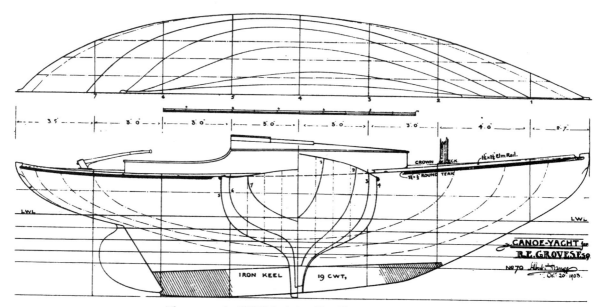

The canoe-yacht, SHEILA, designed by Albert Strange in 1904.
LOA 25ft LWL 19.5ft Beam 6.75ft Draught 3.42ft Sail Area 330sq.ft

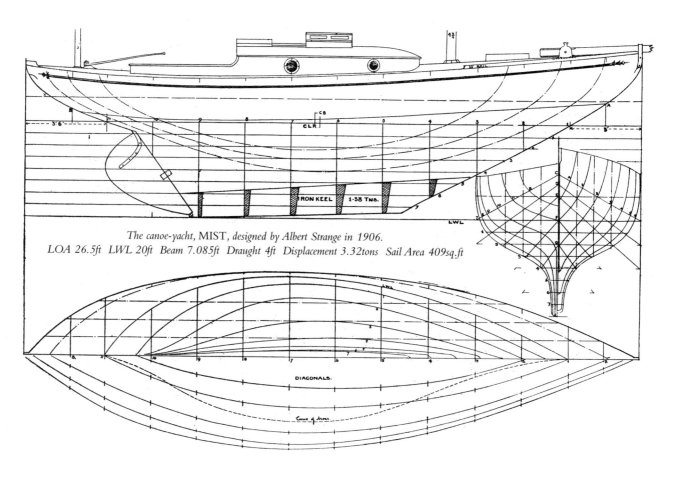

The canoe-yacht, MIST, designed by Albert Strange in 1906.
LOA 26.5ft LWL 20ft Beam 7.085ft Draught 4ft Displacement 3.32tons Sail Area 409sq.ft

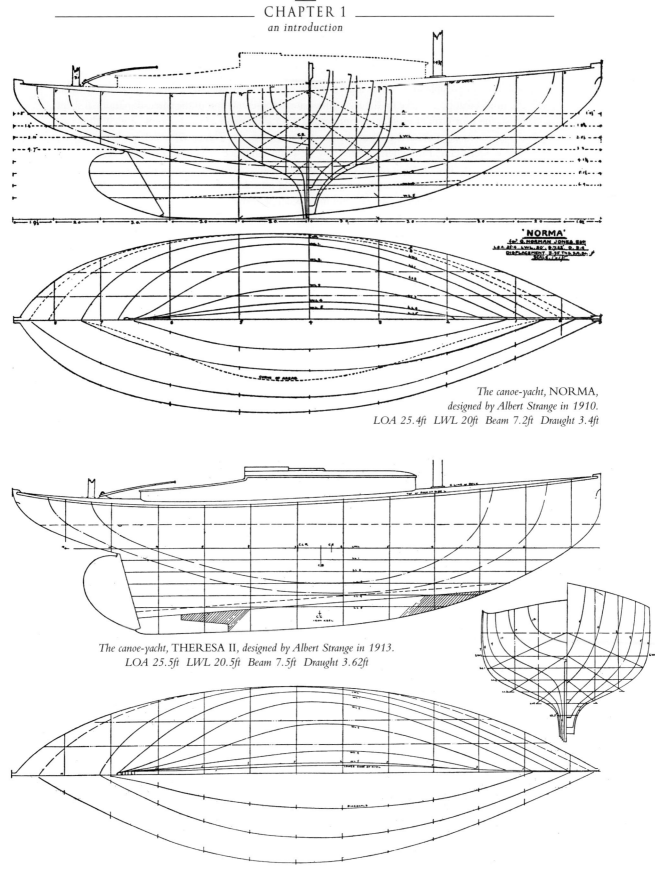

*The canoe-yacht, NORMA,
designed by Albert Strange in 1910.
LOA 25.4ft LWL 20ft Beam 7.2ft Draught 3.4ft*

*The canoe-yacht, THERESA II, designed by Albert Strange in 1913.
LOA 25.5ft LWL 20.5ft Beam 7.5ft Draught 3.62ft*

Profile

Although it is not possible to view one aspect of a lines plan in isolation, since each is inextricably linked with the others that go to make up the two-dimensional representation of the three-dimensional shape, a glance at the four profiles yields some interesting points. From an aesthetic point of view, they each appear to have nicely balanced ends above the waterline, except *Theresa II*, which perhaps does not appeal to modern taste. But, whereas there may be some influence of fashion here, this very snubbed bow profile is actually a function of her rather plumb topsides (a consequence of higher freeboard) and harder bilge; to obtain a more drawn-out profile would introduce flare into the bow, as demonstrated in the majority of contemporary designs and indeed clipper bows.

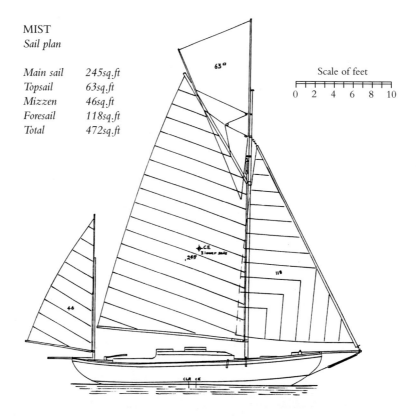

MIST
Sail plan

Main sail	*245sq.ft*
Topsail	*63sq.ft*
Mizzen	*46sq.ft*
Foresail	*118sq.ft*
Total	*472sq.ft*

Scale of feet

0   2   4   6   8   10

Mist has the most drawn-out ends of the four and this, among other things, gives her a relatively long base for the sail plan, allowing a larger area. The heavier *Theresa II*, with her shorter overall length, is by comparison under-canvassed.

Underwater, both *Norma* and *Theresa II* have one long sweep from stem head to heel, *Sheila* has a slight knuckle, and *Mist* a strong one between forefoot and keel. Aside from ease of taking the ground and slipping (and no designer ever spared a thought for the yard foreman who had to get his boats out of the water and stand them up ashore), there are several factors that come into play here: the most important is the influence on the shape of the bow sections – the more cut away the forefoot, the rounder and shallower these sections become. (This is discussed more fully by Strange on page 32 and Fig. 4 page 31.) This difference is clearly seen by comparing the first two bow sections of *Mist* with those of *Theresa II*. The latter should be less susceptible to pounding, although the gain of depth in the focsle is probably what Strange was aiming for here. Other considerations are the area of the lateral plane and its centre of resistance. The difference in area for these four designs is insignificant, but the centre of lateral resistance must tie in with the sail plan.

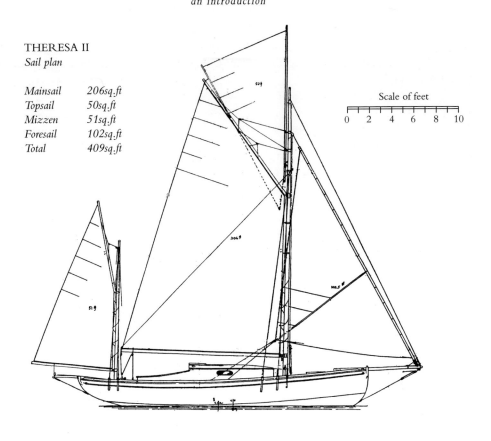

THERESA II
Sail plan

Mainsail	206sq.ft
Topsail	50sq.ft
Mizzen	51sq.ft
Foresail	102sq.ft
Total	409sq.ft

Scale of feet

0 2 4 6 8 10

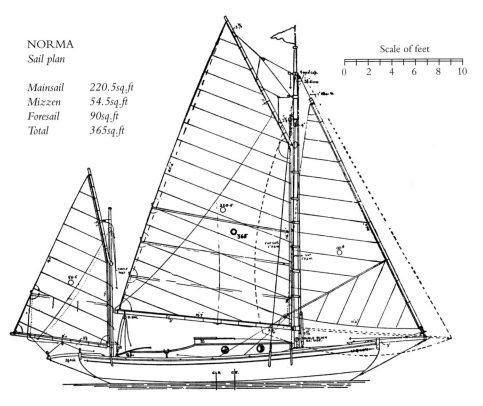

NORMA
Sail plan

Mainsail	220.5sq.ft
Mizzen	54.5sq.ft
Foresail	90sq.ft
Total	365sq.ft

Scale of feet

0 2 4 6 8 10

In construction, the long, fairly straight run of *Mist*'s forefoot would present few problems with timber, whereas finding stock of the right sweep, or jointing shorter lengths, for *Theresa's* forefoot, might prove more costly.

Midsection

Sheila has the most pronounced 'wineglass' section of the four. Her mere 6in less in waterline length than *Mist* and *Norma* belies her much smaller size, which is more accurately indicated by her displacement. Her lines have been described as 'delightfully slippery' and, particularly in light weather, she should prove to be the fastest of the four.

Norma and *Mist* are very similar in midsection, although *Mist*'s greater draught makes her section appear finer. It is possible that her more drawn-out ends will give her a greater effective waterline length than *Norma* when heeled.

Theresa's full midsection is immediately apparent. The beam is carried well down, with a firm turn of bilge. This is also reflected in her deep and more steeply rising buttock lines.

Accommodation

The difference in accommodation between *Sheila* and *Theresa II* is scarcely imaginable from a study of the lines. Although an extra 9in beam gives far more room than one would expect, it is again the displacement that gives the best indication, together with the comparison of the midsections. At her estimated displacement, *Theresa II* is almost 50% larger than *Sheila*. In terms of accommodation, her fuller section allows a lower and wider cabin sole and correspondingly lower and more widely separated settee position. Sitting in *Sheila*'s cabin, one either sits upright on the edge of the settee, with headroom

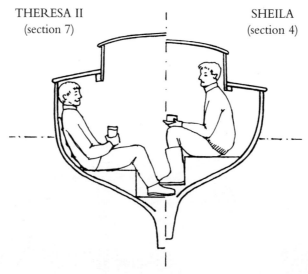

THERESA II
(section 7)

SHEILA
(section 4)

Sketch illustrating the difference in accommodation between THERESA II *and* SHEILA.

under the coachroof but back unsupported, or one ducks under the side deck to lean back against the hull side – a choice has to be made, so to speak, between a stiff back and a stiff neck. On *Theresa II*, Strange has achieved full sitting headroom underneath the side decks and a significantly wider sole (see diagram).

The cost of this considerable extra comfort obviously has to be paid for in other areas – how dearly, is a measure of the designer's success. The heavier, bulkier boat will need more canvas to drive her. She is likely to be slower in light airs and to have a quite different motion at sea. She may also be less weatherly, although again, the skill of the designer is proved if this turns out not to be the case. His job is a highly complex one, juggling an almost infinite number of variables.

A week's cruise in each boat would go some way to evaluating their relative strengths and weaknesses. Their intended cruising grounds would have to be taken into consideration: *Sheila* was built in the Isle of Man for use on the west coast of Scotland by the artist, Robert Groves, who sailed single-handed for much of the time; *Mist* was built in Belfast and is known to have cruised extensively in the Irish Sea; *Norma* was built at Birkenhead for use in the Mersey area and thus required moderate draught; *Theresa II* was built at Leigh-on-Sea in Essex and had the shoals and short, steep seas of the Thames Estuary to contend with.

One criterion likely to be overlooked by any contemporary evaluation of designs of this period relates to the fact that these yachts had no auxiliary power, other than sweeps. Faced with an evening's flat calm in a deep Scottish sea loch for instance, the single-hander can only get his rest once his yacht is in water shallow enough to anchor. In the Thames Estuary, where the channels are narrow and relatively shallow, this is not such an important consideration. In his letter to *The Field* (28th February, 1891), Strange wrote of *Cherub*, '... I must confess that I should prefer an altogether lighter and smaller type of boat for this [single-handed] work as the exertion of rowing 3½ tons of boat and spars against a smart breeze in harbours where there is no room to sail leaves me, at least, rather exhausted after about half an hour'. For many miles north and south of Scarborough, the coast affords little shelter except in such harbours. In the Western Highlands of Scotland, the owner of *Sheila* might well trade some internal

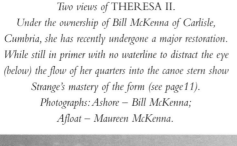

Two views of THERESA II.
Under the ownership of Bill McKenna of Carlisle,
Cumbria, she has recently undergone a major restoration.
While still in primer with no waterline to distract the eye
(below) the flow of her quarters into the canoe stern show
Strange's mastery of the form (see page 11).
Photographs: Ashore – Bill McKenna;
Afloat – Maureen McKenna.

GEORGE F.
HOLMES.

*The last half hour of flood
& no wind.*

YM April 1911

volume and displacement for a yacht of easily-driven form that was lighter work under sweeps. At the conclusion of 'A Sketching Cruise on the Irish Coast' (see JL page 113) Strange relates vividly the back-breaking business of working the 12-ton yawl *Quest* some eight miles under sweeps, round the north-east corner of the Isle of Arran into Loch Ranza.

In conclusion, we have here four outwardly similar boats. Their design not only reflects the different requirements and tastes of four owners in varied cruising grounds, but also, in the differences between them – some subtle, some marked – the hand of a designer who was in full command of the juggling exercise, the balance of one element with another, to produce the successful compromise which the design of every cruising yacht entails.

The four are, quite strikingly, variations on a theme; a theme which Strange had himself developed through long, first-hand experience and a theme which resulted in many a cruising yachtsman treading a path to his door.

THE YAWL RIG

It is perhaps appropriate here to include a brief discussion of the yawl, a rig favoured by Strange for so many of his designs; in fact Strange was referred to as 'The Father of the Yawl' by Maurice Griffiths, one of Herbert Reiach's successors as Editor of *The Yachting Monthly*, in an article reviewing Strange's life and work, published in April 1949.

The rig has always had its admirers and detractors, both from aesthetic and practical standpoints, but at this end of the 20th century, with powerful auxiliary engines, short booms and winch-driven slab-reefing, it is easy to misunderstand why the rig was so popular in the first decades of the century and, indeed, why it subsequently fell from favour. This was certainly not due to any lack of understanding then, or increase in understanding now, of sailing aerodynamics. It is amply clear from Strange's writings included in this book, that neither he nor, I doubt, any other designer, was under any illusion regarding the inferior power and aerodynamic efficiency of the yawl as compared with cutter and sloop rigs. The point was forcefully put by the owner of *Mist* (see page 10) who wrote, in 1916, having converted her from yawl to cutter,

I did a good deal of sailing in Mist *in her original rig. She is a perfect sea boat .... But when it comes to turning to windward, a yawl is a heartbreaking thing.*

However, the choice of a yawl over cutter or sloop, was a choice made on entirely different and far more justifiable grounds in the first 20 or 30 years of this century, than today. Douglas Birt, writing in 1951, commented that 'there seems little excuse for the small yawl, unless it is a sentimental attachment to the type and in particular to the lovely little yawls of Albert Strange'. If we are comparing modern rigs, we can accept this statement and Strange the compliment, but a glance at any of the sloop or cutter sail plans in the present volume will show how different in proportion these were in Strange's day from the modern equivalent. Design No. 119 (page 163) is a convenient example, since it shows alternative sloop and yawl rigs.

Even with a counter stern, the sloop's boom extends right to the taffrail. Transom and canoe-stern designs generally show a boom with substantial overhang. Such a distribution of the sail plan leads to problems in two areas; sail-handling and manoeuvring. Before looking at these in more detail, it is worth noting that, for some sacrifice in performance (but much less than in a ketch), the yawl not only largely overcomes these disadvantages, but has, instead, very real additional advantages.

With regard to sail-handling, the cutter (and for the present purpose I include the sloop) has a relatively large mainsail with longer foot and heavier boom. Anyone who has been shipmates with a boom of, say, some 28ft in length on a 40ft cutter, will know what a lonely and dangerous place her after-deck can be, with a reef half hauled down and a bight of wet canvas nipped in the reef comb sheave by the pennant. In addition, the long foot of the sail demands careful tying of the reef points if the reefed sail is to set well. This is a difficult job if there is any weight of wind in the sail, which there is likely to be, since the cutter relies on at least some wind in the leech to balance her fore canvas. When further reefs are taken, the long-footed gaff sail becomes progressively less efficient as its leading edge is reduced in length.

The relatively tall mainsail of the yawl does not suffer from this problem to the same degree. Her head can be held up into the wind while reefing, by sheeting the mizzen hard and, of course, the shorter boom is well inboard where the deck is wider. A well-proportioned yawl rig on a reasonably fine and light displacement hull will often sail quite well without her mainsail and this can be a very useful and comfortable rig at sea in strong winds or passing squalls.

With regard to manoeuvring, the yawl comes into her own. It is a sad fact that manoeuvring under sail is a dying art, due in large part to the reliability and power of the modern auxiliary, and to the encouragement, even necessity, of its use imposed on us by the modern marina and congested mooring areas. Manoeuvring under sail when there is no recourse to auxiliary power even *in extremis*, produces fundamentally different criteria which it is difficult for the modern yactsman, however little he uses his auxiliary, to appreciate.

Particularly when short-handed, one of the most desirable, and most difficult, things to achieve in a small yacht when manoeuvring is the combination of slow speed and full control. A cutter in many situations will not handle without her mainsail, yet with this sail set she is powered by the greater proportion of her working canvas. A good yawl has the right proportion of sail area in her mizzen and headsails to handle quite precisely at slow speeds. The little yawl, *Sheila* for example (see page 10) can be put about and, without touching a sheet, will lie to with the helm untended, making virtually no way while the mainsail is lowered and stowed. From this position, as soon as the weather jib sheet is released, she will come back through the wind onto any desired course; something I did not believe until I tried it myself.

Not all yawls will do this, but the mizzen will at least stop the head paying off while the mainsail is being lowered and it is then a simple matter to lower the mizzen even when, for example, running to a berth down wind.

A mizzen can, of course, also prove very useful in getting away from an awkward berth, as the following incident described by Strange demonstrates. He was aboard his friend's 33ft-LWL yawl *Quest* (see page 125), lying to two anchors in Dundrum, on the east coast of Ireland, when a heavy NW gale drove the yacht ashore just as the tide began to make.

We bolted on deck, hardly able to see through the mixture of rain, spray, and sand which the wind tore up from land and sea, and as the tide was flowing we decided to try to get the shelter of the pier half a mile to windward. There was no difficulty in getting the anchors, as both of them were huge bunches of weed, and with reefed foresail and mizen [sic] only we positively tore through the water with the strong flood under us and shot up into the wind alongside an old dredger, where there happened to be just water for us. This was another occasion when the yawl-rig was a godsend, for there was so much wind that I doubt if the yacht would have carried even a reefed trysail. As it was, she lay down to deck edge under not much more than 150sq.ft of sail, yet she handled splendidly.

The key to a successful yawl is that the hull should be designed with the rig in mind. H J Suffling, whose design for a yawl appeared in the January 1917 number of *The Yachting Monthly*, puts the point well,

VENTURE
owned by Ian Taggart.
The bermudan-rig
conversion was designed
by Walter Easton in
1937 for Pat Walsh,
who had previously
owned SHEILA and
SHEILA II.

Photograph:
Frontispiece to
The Single-Handed
Yachtsman,
Francis B Cooke, 1946.

My first requirement is for a safe, though small, single-handed cruiser. This, I consider, demands the yawl rig, and then a suitable hull to suit this sail plan. My idea is that a yawl should be longer on the waterline in proportion to her beam than is usual in small boats.

The dimensions which I have decided on after much thought and many trials are: LOA, 28ft; LWL, 22ft; Beam, 6ft 8in; Draught, 4ft 3in; Displacement, 4.38 tons; Sail Area, 370sq. ft.

These measurements ... produce a boat of fairly liberal displacement, yet the lines denote an easily-driven hull... a beamier boat is not so suitable for the yawl rig, because it is impossible to obtain the necessary sail area without the use of long and lofty spars, which are undesirable for our purpose. The narrower yacht responds more quickly to smaller sails, and in this respect is well suited by the yawl rig. In my opinion the single-handed boat which is intended for long passages must be a yawl. With this rig it is easier to balance the sails than with a cutter or sloop, and the advantage obtained by lowering the mizen when sailing free in a strong wind with considerable sea, in contrast to running with a reefed mainsail in a cutter or sloop, has only to be experienced once to be appreciated for all time. This opinion regarding the yawl rig is held by Mr Albert Strange, whose views regarding the design of small craft must be regarded as almost final.

Suffling sent his draft lines to Strange who adjusted and refined them just before his death. The yacht, named *Venture* (see JL pages 142-3), was built at Lowestoft in 1919 and proved a great success. (She is now based on the Clyde and is cruised extensively.) In 1922, now needing a larger boat, Suffling's brother N R Suffling wrote, 'We were so surprised at the fine sea qualities of little *Venture* that I decided I could not do better than build from the same design, increasing the size to 6 tons'. This yacht was named *Charm* and was 33ft LOA. The exercise was repeated twice more, with *Charm II* of 40ft LOA and *Sea Harmony*, 33ft LOA. To an account of his successful maiden cruise from Lowestoft to Falmouth and back in *Charm*, N R Suffling (in the same forthright style as his brother), added a few lines in defence of the yawl rig.

I see one writer who owns a fine cutter with a motor recommending his rig in preference to the yawl. I quite agree that if you have a motor it does not matter if you have one mast, or two, or none!

CHAPTER 2

The Design and Construction of Small Cruising Yachts

*The Yachting Monthly, August 1914 to April 1915
and October 1917*

SHEILA H.Y.C
UNDER FULL SAIL

ROBERT. E. GROVES. 1910

YM Sept 1910

THE DESIGN AND CONSTRUCTION
OF SMALL CRUISING YACHTS

Geo. Holmes   *YM Sept 1916*

On Thursday evening, 14th September 1916, having just returned to Scarborough from a brief visit to London, Albert Strange wrote a short letter to his good friend, Harrison Butler, another talented amateur designer. It was chiefly concerned with the scantlings of the design (later to be named *Venture)* for H J Suffling of Great Yarmouth, but went on to say,

> *I saw Reiach in town for a few moments, but we did not get to business, so I suppose the idea of publishing my articles on yacht design, with additions and corrections, is 'off' for the present anyhow. I expect there will be difficulties over it even when things become normal — but I retain copyright and so far am on the right side.*

Tragically, the war and failing health, and possibly some difference of opinion with Herbert Reiach, editor of *The Yachting Monthly*, prevented publication of these articles in amplified and corrected form. Nevertheless, the text as it originally appeared in the magazine is eminently worthy of reproduction. Not least among Strange's talents was his ability to write lucidly and elegantly on whatever topic was under consideration. His exposition of the fundamental process of yacht design, within the parameters he has set, is one of the clearest and most readable texts available. From an historical point of view, it also tells us much about the thinking which went into his own designs, and makes fascinating reading for anyone who owns, or maintains a yacht of this period.

We cannot know how Strange himself would have revised his work, but we have ventured to correct one or two typographic errors, and add footnotes where it was felt some explanation of the text was necessary.

The Design and Construction of Small Cruising Yachts

BY

ALBERT STRANGE

[TOOLS OF THE TRADE]

NOTWITHSTANDING the numerous works on Yacht Designing and Naval Architecture which have appeared in England and abroad during the last thirty years, there is yet a greater amount of mystery surrounding the design of small yachts than should be the case considering the very large number of yachtsmen we fortunately possess. Perhaps this may be due in some measure to the fact that most of the works in question, are in many instances, of too advanced a nature, and cover too great a mathematical scope for the ordinary amateur yachtsman to assimilate easily. It is a sad fact that the average man forgets his school mathematics very rapidly unless he enters a profession in which he is compelled to make use of the science every day. And it is unfortunately true that hitherto mathematics, as taught in our schools, does not enter the region of practical work soon enough to become interesting and useful to the average mind.

Yet any man with a knowledge of ordinary arithmetic, combined with a fairly good eye for form, and, what is of greater importance, who also possesses a practical knowledge of the requirements and behaviour of an ordinary yacht, should be able to place his ideas on paper in an intelligible and practical form, so that from his drawings a yacht could be constructed by the average builder, which should be satisfactory both as a habitation and as a sailing vessel.

It is indisputable that every yacht worthy of the name should be thought out carefully before construction is commenced. But unless the necessary plans and the few simple calculations required to ascertain her displacement, her centres of buoyancy and lateral resistance are made, it cannot be said that she has been thought out. A certain amount of mystery, chance, and experiment fortunately cling round even the most carefully designed vessel. But the undesigned boat, though she may be considered a success, remains a mystery to the end. There are no facts of real importance known to anybody so far as her elements are concerned, and it is generally the case that an unsuccessful boat, though her faults are easily explained by the expert, cannot be altered in any material way, unless she is completely rebuilt.

Therefore, feeling sure that the study of the elemetary principles of naval architecture will be found very fascinating if once taken up, and that its difficulties are greatly overestimated by the uninitiated. I venture to commence this series of papers, intended solely to help the beginner, in which I shall not touch on any abstract principles or reasoning except

---

* This article was written by the late Mr. Albert Strange as an introduction to the series of essays which he wrote in 1914-1915. We differed on one or two of the points touched upon, and the author, conscientious and courteous as he ever was, insisted on its being held over till we had opportunity for discussion. That opportunity, alas, will never be, so we publish the essay as written.—ED.

those absolutely necessary for the real understanding of the subject. I shall confine myself solely to the most elementary facts, rules and formulæ, explaining them in as simple and plain a manner as I can command; presuming only that my readers understand what a scale drawing is, and what relation it bears to the real yacht, and are also able to add, multiply, divide and subtract decimally. The necessary tools are not numerous or expensive, and when once obtained will, with care, last for ever. These are:

ference in size producing endless combinations, available for a dinghy's sections or for the sections of a 19-metre yacht.

A Long Flat Scale, divided *decimally* into scales of 1″, $\frac{3}{4}$″, $\frac{1}{2}$″, $\frac{3}{8}$″, $\frac{1}{4}$″; this should be at least 24 in. long.

A Set of ordinary Cardboard Scales, in a case, price 1/-; useful for cabin and other plans when scales of feet and inches are wanted.

Splines and Battens.

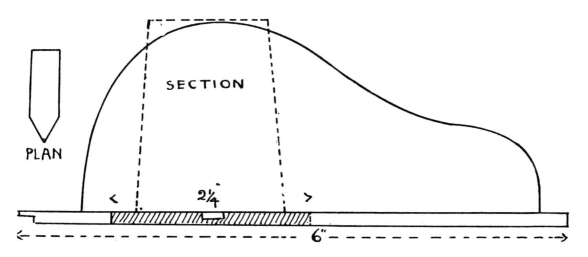

Fig. 1—4lb. LEAD WEIGHT. ORDINARY SHAPE (Full Size)

A Drawing Board, 31″ by 22″ or larger, perfectly rectangular, flat and true. The best kind have a strip of ebony in the left hand edge, with a T square 31″ long. A smaller Board, 24″ by 16″ is also very useful.

A set of mathematical instruments of good make, not necessarily highly expensive, though the best are the cheapest in the long run.

Two Set Squares, preferably of celluloid, 45°; one 6″ edge, one 12″ edge.

A set of Harrison's Graded Scale Curves, 5″, 7″, 10¼″, 13¾″, in celluloid or pearwood. With these four curves any conceivable curve can be drawn and are all the curves that are generally required. They are all shaped as shown, the difference in size producing endless com-

Drawing Paper, Tracing Paper, Indelible Indian Ink, Pencils, etc.

Splines and Battens are usually made of lancewood of the best quality. They are also made of celluloid. For the beginner the lancewood battens are best, as they have more backbone than celluloid ones, and will not easily bend to an unfair curve, which those of celluloid will easily do. A very yielding batten is very misleading to the inexperienced draughtsman, and at first as stiff a batten as can be held in place by all the weights should be made use of until the eye is trained to accuracy in judging what is a fair curve.

My most useful battens are 4 ft. 6 ins. long, one is three-sixteenths square in

the centre, gradually tapered to one-eighth square at one end, and a trifle less at the other. This is a stiff batten, useful

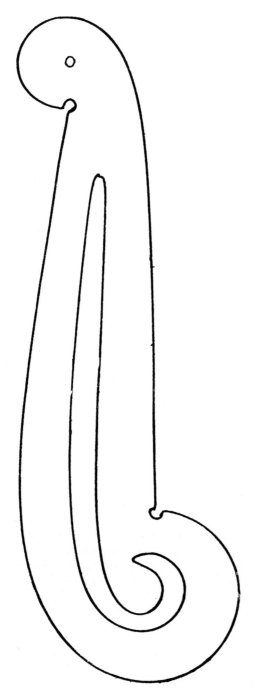

Fig. 2—7 inch (Full Size)

for sheer lines, etc. Another, perhaps the most precious, is 4 ft. long. One-eight inch square at one end, gradually

and evenly tapering to one-eighth inch by one-fortieth at the other. It will bend into a very short, round curve, fair or unfair at this end. Other battens are similar in shape and length but of varying degrees of stoutness.

The long Ship or Yacht Curves are not necessary with battens of the type explained (Fig. 3). The use of these long curves is very apt to produce a mechanical sameness of shape not desirable to encourage in a beginner.

The best type of weight is that invented by the Editor of the *Yachting Monthly*, (M.I.N.A.). They can be purchased from Messrs. W. F. Stanley and Co. In Fig. 1, however, I show the older form of weight, one which can be easily made.

The best paper to use is Hand-made Hot Pressed, either " Imperial " or " Double Elephant " size, according to the size of drawing board used. Imperial size is 30 ins. by 22 ins.; Double Elephant is larger, 40 ins. by 27 ins.

This paper is far harder and smoother than the machine-made paper usually sold in rolls, and will stand the use of indiarubber to a far greater degree without losing its surface. Good " outsides," that is, paper not of the most perfect finish, costs 3/9 per quire Imperial; Double Elephant size, 6/6 per quire.

There is yet one instrument that it should be the ambition of every budding yacht designer to possess, and that is a Planimeter, or Area Measurer. The cheapest kind measures square miles [1] only, and costs £2 2s. The next size will measure areas to various scales, and costs £2 10s. 6d. Either will save an enormous amount of time and ensure accuracy. They are not difficult to use, and, of course, are not indispensible. Naval Architects had to work without them not so many years ago, and managed to obtain accurate results by ordinary calculation in the more leisurely days of the past.

It is not desirable to make your plans more than about 40 ins. long. The eye cannot take in a larger drawing than this, unless it is viewed in an upright position from a distance of 5 or 6 feet. From this

1 'square miles' should be square inches

point of view a line that may seem quite fair at close quarters will soon show any unfairness, or " lumps," particularly in sheer lines or water lines.

Tracing paper is of two kinds. One is highly transparent and glazed, and is used for making tracings of the pencil drawings. It has a slightly greasy surface sometimes, which repels the drawing ink. This bad quality can be removed by rubbing the tracing paper over with a little powdered French chalk. This paper is best obtained in rolls of 20 ins. wide, and should be of fairly stout quality. From an ink tracing any number of prints can be obtained in black line on

stouter, and far more durable. It is also more expensive.

Designs should not be inked in, but should be drawn with a firm clear pencil line and traced when completed. But all *construction* lines should be inked in first of all on the pencil drawing as they will not then be lost in making the inevitable alterations which will constantly occur during the progress of the design. (Fig. 5). These constant alterations and corrections are somewhat disheartening at first, but as facility is gained there will be fewer to make. They can never be entirely avoided, and I have never yet met the designer who can make a complete

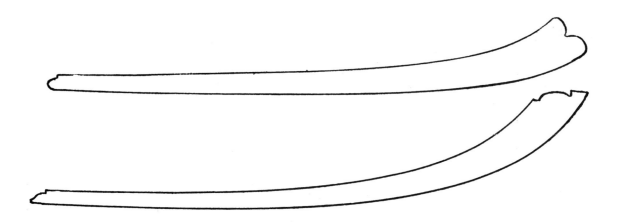

Fig. 3—CURVES 24 inches LONG

white ground at a small expense, or " blue prints " can be also made from it.

The other sort is a semi-transparent unglazed paper of bank note quality and thickness. It is very useful for placing over the sheer and half-breadth plans in order to make the cabin plans. It is sufficiently transparent to permit of lead pencil lines being seen through, and its use saves a great deal of time and ensures accuracy. It takes ink well, and does not lose its surface when india-rubber is used, but it is too thin to stand scratching out with a penknife.

Tracing cloth is similar to the glazed transparent tracing paper, but is much

design without alterations, and I do not believe that he exists.

Tracing is a wearisome occupation and requires a good deal of practice before good work can be produced. But it is better to trace from the original drawings than merely to ink these in, as if your design is to be built from, the builder and the sailmaker and rigger will want plans and you should have copies too for reference in case any accident should occur to the original.

The most useful pencils are those of " F " and " H " grades. " HB " is also useful. Do not use cheap, common pencils as they will not rub out and you cannot keep a good point on them. See

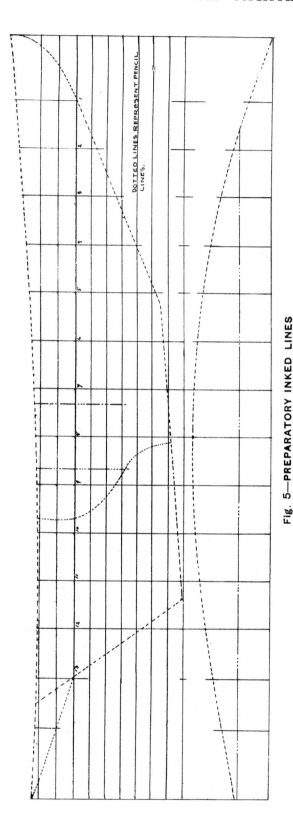

DOTTED LINES REPRESENT PENCIL LINES.

Fig. 5—PREPARATORY INKED LINES

that they are hexagonal, or square, in section, as the round ones roll off the board and break their points. A small piece of glass paper will enable you to renew a blunt point by a few rubs, but you will have to use a *sharp* penknife to remove the wood occasionally.

Should blots or smears occur on tracing paper they should be allowed to dry and then be scratched out with a very sharp penknife, taking care to scrape one way only (not backwards and forwards). The hard ink-eraser used by typists is very useful in removing thin smears sometimes caused by the ink from the pen running under a batten or set square when the pen is too full, or held too close to the batten.

The amateur draughtsman will probably try to use makeshifts in the way of weights or battens. I have known a beginner to use books, clothes irons, inkpots and what not for weights, and the ends of fishing rods for battens, to his great discomfiture. It is a fact that you cannot produce good work with improper instruments, and one of the very first essentials is to cultivate the most extreme accuracy in measurements and drawings, if the result is to be a success. Remember that every error you make in a drawing to the scale of one inch to a foot is magnified twelve times on the actual vessel. Even in making a preliminary sketch it is most essential that accuracy of measurement be observed. In no kind of designing is accuracy more desirable or more difficult to obtain than in the design for a yacht, and even with the greatest care an error will creep in and pass undiscovered until the builder lays the lines down on the loft floor to full size. And no beginner can have any conception of the extent to which an error will reproduce itself all over the vessel until he has learned by bitter experience how necessary extreme care is, and what a horrid nuisance a slight mistake of a small point may prove to be. Therefore let the careful man be still more careful, and the careless man mend his ways, when he takes his work of designing in hand. There is a moral gain in it to him that is worth striving for to the utmost.

The Design and Construction of Small Cruising Yachts.

BY

ALBERT STRANGE.

I.—FORM.

BEFORE entering on the operation of actually designing a yacht it may be well to consider the question of form and its accompanying qualities.

A cruising yacht should be easy in a seaway, of moderate draught and of good initial stability. She should have great righting power at any angle of heel, from five to fifty degrees. She should also be " handy "—that is, she should obey her helm quickly and carry her way well round in tacking—be capable of being hove-to unattended in moderate weather, and have a good turn of speed without being over-canvassed. She should be " dry " in a seaway—that is, not take heavy water on board over her bows, afford good habitation below decks according to her tonnage, and her construction should be sound, strong and efficient, without undue weight.

Some of these qualities are mutually contradictory, and it is in the skilful adjustment of these contradictions that the ability of the designer is shown. For instance, habitability means displacement, displacement means weight to be driven. If it is to be driven with moderate sail-area it must be well disposed longitudinally, and, however well disposed, the cruiser can never successfully compete with a racing boat of similar size and sail-area unless the racer be exceptionally slow.

Ease in a sea-way largely depends upon the shape of midsection and arrangement of ballast. The midsection practically governs the whole body shape from stem to stern, and consequently the relative positions of the Centres of Gravity (C.G.) and of Buoyancy (C.B.). If the C.G. is

relatively low and the C.B. high, the yacht will have great metacentric height, and be quick and jerky in her motion among waves. She will be very stiff initially, i.e., will resist heeling at small angles, and consequently very quick to return from any inclination to the upright. Here again a contradiction—a very stiff yacht is not always an easy sea-boat.

The shape of the midsection again depends upon the size of the yacht. The form of midsection suitable for a yacht 20ft. long may not be the best for one 40ft. long, and still less so for another of 60ft. length. (Fig. 1.) In a very small yacht initial stiffness and ability to carry sail cannot be sacrificed to easy motion if she is to be comfortable to live on, as a small yacht that is without good initial stability is far from being a comfortable ship, though she may be quite safe.

Then the two qualities of handiness and the capability of being hove-to are to some extent contradictory. A very handy and quickly turning boat should not have too long a keel, and a boat that is intended to be hove-to should not have a short keel. For racing purposes a boat can hardly be too quick on her helm, but in cruising this quickness, entailing continual watching, is undesirable. A much cut-away hull may perhaps be hove-to under very favourable circumstances, but she is then so apt to fore-reach fast that she may as well be sailed. A cruiser must be able to look after herself at times—a racer rarely is.

It is often asserted that every improvement in cruising yachts has been initiated in the search for speed induced by racing. This is hardly exact, and its truth depends entirely upon the type ultimately evolved

27

by the rule under which they are constructed. So far every rule has "run to seed" in a bad direction. Some rules are so bad initially that they soon bring about their own downfall, and any rule, however good it may seem at birth, will be certain to be so manipulated by clever designers that speed will be obtained at the expense of all other qualities. Nearly all rules begin by producing desirable boats (a cheering assurance of the continuing sanity of

than the other assertion, for it is first of all necessary to produce a single type of excellence in fishing boats. But as a matter of fact fishing boats vary even more in shape than do racing yachts; there is no fixed type of fishing boat. They vary as the coast, the character of the fishermen, and the nature of the fishing compel them to. In one thing only do they all agree, and that is that each type is the product of a certain set of circumstances, and,

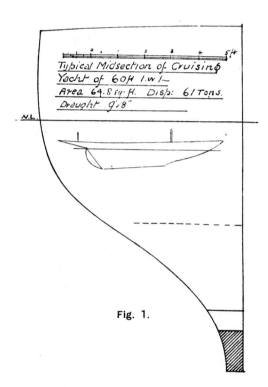

Typical Midsection of Cruising Yacht of 60ft l.w.l.
Area 64.8 sq. ft. Disp: 61 Tons.
Draught 9'8"

Fig. 1.

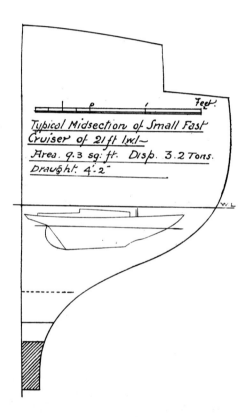

Typical Midsection of Small Fast Cruiser of 21ft l.w.l.
Area. 9.3 sq. ft. Disp. 3.2 Tons.
Draught. 4'-2"

the human mind), but all end by so modifying the desirable boat that she becomes undesirable as a cruiser. Excessive complication and excessive simplicity of rule produce in the long run about the same evils, and whether it is possible to devise a rule that will give "freedom of design," and yet give a form of hull quite desirable for cruising, is to my mind a doubtful question. At any rate it has not yet been accomplished.

On the other hand, there are those who stoutly assert that the best type of boat as a sea boat (or, in other words, a cruising yacht) is pointed out by the forms of fishing boats. Perhaps this is a greater fallacy

given the circumstances, is the best type discoverable. Now this really brings us back to our first conditions, that the good cruising yacht should be within the limits of her size fast, handy, safe and habitable; for these are the circumstances in her case that she should be adapted to. The limits of her *size* are fixed by the nature of the waters upon which her cruising will be done.

Transversely, that is, in midship section, the *easiest* form approximates to Fig. 2 (A), but unless the beam is moderate in proportion to length the displacement is disproportionately heavy. This form has not great initial stability, and is therefore un-suitable for very small yachts; yet its ulti-mate stability is unlimited, being derived from weight low down, *i.e.,* a low centre of gravity.

Fig. 2 (B) shows a form of great initial stability, and is one in which the displace-ment is not so great in proportion to length on L.W.L. It is suitable for small centre-board yachts, but it is not easy in a sea. But it is more easily driven at high speeds than A, and if properly ballasted is safe if no water gets below. But at great angles of heel the area of its load water plane

good deal of attention from designers, notably from Mr. Charles Nicholson. Lifting means loss of sailing length, and it is probable that the pronounced tumble home of the topsides of his recent creations has been adopted to modify this lifting tendency, as well as to gain other things conducive to speed. It is not a new device, our old sailing men-of-war possessed it in a marked degree, the main reason in their case being to bring the weight of the guns inboard, and thus prevent excessive rolling. It was seen also in some of the older types of American yachts. In a cruiser there is

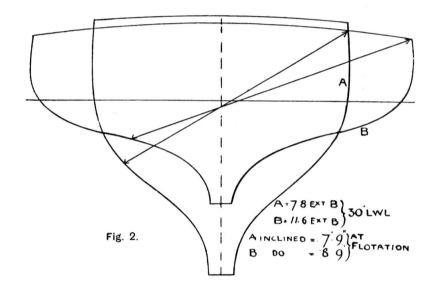

A = 7 8 EXT B ⎫
B = 11.6 EXT B ⎭ 30′ LWL

A INCLINED = 7′ 9″ ⎫ AT
B DO = 8 9 ⎭ FLOTATION

Fig. 2.

diminishes much more rapidly than that of A, and at ninety degrees is only about two-thirds, and the weight on the righting lever is much less at all angles. Although ninety degrees is not an angle that need be taken into consideration, it should not be for-gotten that a wide and somewhat shallow form of hull diminishes its area of flotation, upon which stability in part depends, very rapidly almost directly the nearly upright position is departed from.

With vessels of normal type it is almost impossible to produce a design that will not lift bodily on being heeled, and the shallow type lifts sooner than the deep, the part of the hull put in to leeward by the action of the wind on the sails being greater than that taken out. In modern racing yachts this feature has received a

no particular advantage in it, unless the yacht is of exceptional size.

It will thus be seen that when all the requirements of a cruising yacht are taken into consideration there is a very small pos-sibility of largely departing from the best type of hull at present in existence. That is to say a type which has a fair amount of underwater body, a firm bilge, good freeboard and a proportion of beam vary-ing from more than one-third of the L.W. length in small yachts to about one-quarter in large, and an extreme draught of from one-quarter to one-seventh of the load water-line—according to the length of the yacht. (Fig. 3.)

But it does not follow that because these proportions are to some extent stereotyped that there is no room for modifications and refinements both in hull form and accommodation, as well as the variations required by personal needs and local physical conditions. It is on these points, and in sail plans that there is still room for diversity of treatment in theory and practice, and it is in solving the many conflicting requirements of a cruising yacht that the charm of amateur yacht designing is most felt.

The desire to obtain a great length of effective body above water by means of long overhangs forward and aft has often led the amateur designer astray. A moderate amount is of undoubted benefit, as by its

paratively shoal hulls with deep fin or semi-fin keels and long overhangs demand for cruising a very high cabin top and require a large sail area. They are wet and uncomfortable in a really disturbed sea, though fast and stiff in landlocked and non-tidal waters. They are inherently weak in structure, owing to the great length of deck opening without transverse beams to hold the hull in shape.

With regard to the after overhang, given a fairly compact and easy form of midsection, it is not possible to draw out the buttock lines to an undue length if they are

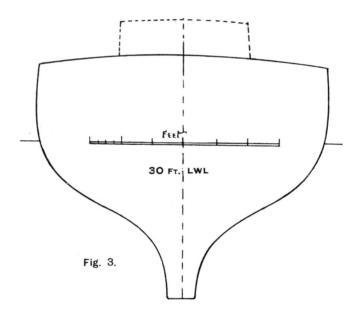

30 FT. LWL

Feet

Fig. 3.

means the bow lines are lengthened and eased when the vessel is heeled and moving through the water and the bow wave resistance lessened. The same good effects are produced by a well-shaped after termination to the hull, but excessive overhang at either end defeats the objects desired.

Excessively long fore overhang seriously decreases the amount of useful room obtainable in the forecastle, as it is impossible to model a fairly full midship section into a long bow overhang and at the same time obtain depth of cabin and forecastle. A shoal form of midsection easily lends itself to long overhangs, but is undesirable when the best cruising qualities are sought. The types common in America of wide and com-

to be fair and easy curves. Any attempt to overdo the natural length fixed by the shape of midsection and the true curvature of the buttock lines, at once produces an ugly hollow curve about the counter, or else compels such a full round at the after end of the L.W.L. that very undesirable wave-making is caused at high speed, together with such an alteration of the centre of buoyancy as to destroy the proper balance between it and the centre of gravity of the hull. Such boats bore badly when pressed under canvas, and in consequence become almost unmanageable.

But it must be remembered that if the form of the waterlines aft is too fine there is a waste of power caused by the undue diminution of the area of the inclined load water plane. This may also cause the yacht to suck down aft to a dangerous extent when travelling at speed. It is also wasteful in another way, as a good deal of the impulse derived from the effort of the "following wave" which is flowing in from the surrounding water as the vessel moves forward is thrown away. The fact is that if a boat is to behave well at speed she should have well-balanced ends, and a centre of buoyancy under all conditions of heel not greatly removed from the centre of the load waterline, or close to the greatest transverse transverse waves, which travel along the body of the vessel. Hence there is not enough to "fill up" if both ends are equal in displacement. The true position of the centre of greatest area should be slightly astern of the centre of the L.W.L., as then a markedly longer and easier bow is obtained, which in itself makes for speed, and the crest of the "following wave" or "filling up wave" is brought further aft, where its power is better utilised.

In bodies totally submerged, such as those of submarines, lead bulbs or fins, and in fishes of nearly all kinds, the forward end

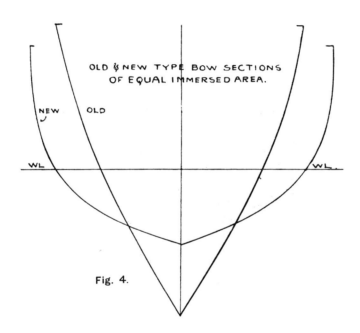

OLD & NEW TYPE BOW SECTIONS OF EQUAL IMMERSED AREA.

NEW   OLD

WL   WL

Fig. 4.

area of section. In most well designed modern vessels it is always slightly aft of this position, whereas formerly it was thought to be an advantage if it came somewhere ahead of the centre of load waterline —hence the old "cod head" and "mackerel tail" style.

It is still sometimes argued that the section of greatest transverse area should be exactly in the centre of the L.W.L., because the amount of water displaced by the bow should completely fill the space occupied by the after-body as the vessel moves through the water. But it must not be forgotten that part of the water displaced by the bow is lost in the "waves of divergence," which form a different set of movements from the is rounder and blunter, and the after end longer and finer than that which long experience and experiment have proved to be the best for *floating* bodies. In these submerged forms it must be remembered that there are no surface waves formed at all, and as the surface waves caused by the vessel's forward movement form practically all the resistance that a moving floating body has to overcome, except the resistance due to skin friction, which is common to both floating and submerged forms, deductions drawn from the shapes of fishes, etc., have no value when applied to an entirely different set of circumstances. So the cod head and mackerel tail theory must be abandoned by us, at any rate.

A comparison of the lines of modern racing yachts with those of racers of twenty or thirty years ago reveals a very marked difference in the shape of the forward waterlines. Formerly, under the 1730, or "plank on edge" rule, and also in the earlier yachts built under its successor, the length and sail area rule, all vessels designed for great speed had fine or even hollow waterlines forward. Now it is quite the reverse. Waterlines are now round forward, and give the unitiated the idea that the excessive fullness of the cod head era is once

obtain on her total length. Not hollow, but at the same time not too full. The hollow is not wanted because we should have some amount of "cut away" under water forward, which naturally cuts off the hollow fore ends of the waterline, besides, a hollow bow is a weak bow in a sailing vessel unless she has very great proportion of breadth to length and a very long keel when some hollow is almost inevitable in the waterlines, especially the lower ones.

It is important to remember that although the plan of a sailing vessel is drawn to re-

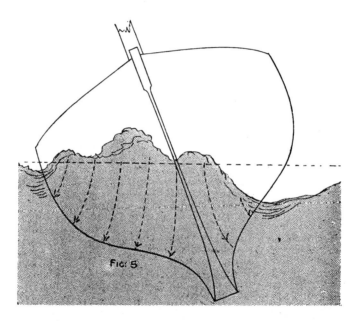

FIG. 5.

more returning. But as a matter of fact for all practical purposes bows are, *in body,* as fine as ever they were. It is only because so much of the underwater body of the bow is now cut away that the lines *have* to be round, in order that there be any bow at all. And such roundness is only good for racers, and in them only good when the yacht is well heeled and moving at a high speed. At low speeds in disturbed water these bows pound and splash most undesirably. They are most essentially not cruising bows on account of this feature and also on account of the enormous sacrifice of internal accommodation in the matter of headroom they demand. (Fig. 4.)

So our comfortable fast cruiser must have as long and easy bow waterlines as we can

present her in an upright position, this attitude is seldom assumed when the vessel is under way. She is always heeled to a greater or less degree, and the consideration of the shape she will assume when heeled at various angles is extremely important. Research in this country, and more particularly in America, has brought to light the path in which particles of water travel when she is moving ahead. These particles pass *under* the boat from bow to stern, and do not so much move diagonally round her sides as was formerly thought. They take a track which is indicated more or less closely by the buttock lines when she is in an upright position, and when she is heeled they pass under her body in paths at nearly

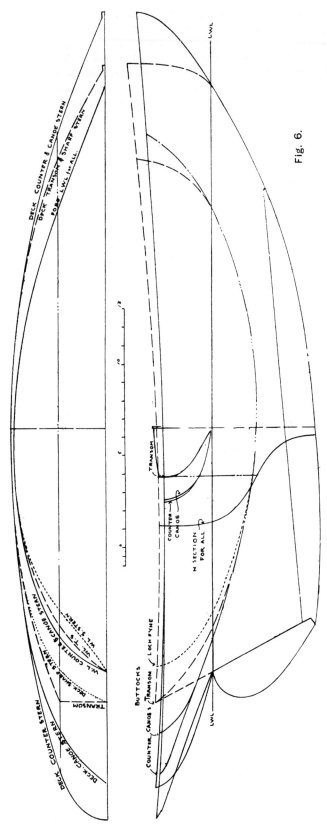

right angles to the inclined waterline.
(Fig. 5.)

In order, then, to properly gauge the
easiest path in which they shall move, the
model should be carefully considered under
an amount of heel that may be considered
consequent upon the application of a certain
wind force. This will be dealt with more
particularly when we arrive at the point of
making the design, but a glance will show
that when a vessel is sailing at a high rate
of speed and consequently heeled to a decided angle from the upright position, the
shape of these paths is of great importance.
At the bow, where the wave formation begins, and abaft the end of the waterline
where the waves end, there is a considerable
amount of resistance. This resistance can
be lessened very largely by lines of easy
curvature, such as can be obtained by overhang forward and aft, and this fact is one
of the explanations of the comparatively
large improvement in speed obtained by
modern yachts on a reach or on a wind as
compared with vessels of an earlier type,
with their straight bows and short tucked-
up counters. Running dead before the wind
this difference is less marked, owing to the
fact that the faster a vessel runs the less
is the propulsive power of the following
wind, though even on this point there is a
great and perceptible difference between
vessels of the same displacement, but having ancient or modern forms above water.
The lengthening out of the diagonals in
modern yachts is a very large factor in the
matter of racing speed as well as in the
matter of stability. The amount of extra or
" surplus " buoyancy inherent in vessels
with large proportional above-water body at
the ends, greatly exceeds the loss due to the
extra weight of these overhangs, and it follows that even in cruising yachts as much
overhang fore and aft as can reasonably be
obtained without loss in other important
points should be a matter of careful consideration.

It is to these facts that the marked in-
feriority of vessels with sharp sterns or
transom sterns, as compared with vessels of similar type and displacement,
but with fair overhang fore and aft,
is due. A sharp-sterned vessel of the
" Loch Fyne " type has much shorter
buttock lines than any other type. A
transom-sterned craft has the advantage

of somewhat longer buttock lines than the sharp-sterned vessel, but there is no difference in the bow lines, and there is also a considerable amount of eddy-making at the tuck if it is much immersed, which, of course, is detrimental to speed. But on the Clyde, where so many small yachts are finished aft with a transom, there is nearly always a considerable amount of immersion, due doubtless to the fact that they have deep 'midship sections, and also to the desire to get as long buttocks as possible, even if the tuck has to be immersed to obtain them.

Fig. 6 shows the difference in the form of the buttock and bow lines, and also in the shape of the after end of the load waterlines between hulls having counter, canoe-stern, square stern and sharp stern. The two latter have markedly shorter buttock lines, although the length of load waterline and the shape of midsection are all alike. A graphic comparison of this nature is far more illuminating than pages of prose, and should make quite clear the value of overhangs, even of the moderate kind here shown, and which are common to cruising yachts.

Itchen Ferry boats, being of a shallower midsection generally, have much higher set tucks, and there is little doubt that such a form of midsection as these boats have is better suited to the transom stern than a deeper bilged section can be. But in any transom or square-sterned boat there is inevitably a good deal of "drag" and eddy making aft, and it is only on account of an actual limit of overall length that there can be much justification for the type. If a yacht has to be, by rule, or circumstance, limited to, say, 25ft. overall length, then it is better to dispense with overhangs, for by so doing you obtain a larger actual boat than if 5ft. of this limited length were placed in the overhangs. In the former case you get at least 24ft. of length on the waterline with its proportional beam well spread out fore and aft, whilst in the latter you only obtain 20ft. with its proportional beam, displacement and accommodation.

This case, of course, presupposes that the sail area is not limited. If it were, the conditions are altered at once. But as the rough apportionment of sail area for cruisers may be taken as the square of the waterline length, in the one case as you get 576 sq. ft. and in the other only 400, and the 576 would, other things being proportionate, be carried longer and better than the 400 sq. ft., and the difference in speed due to difference in length alone would be very considerable.

The true standard of comparison is, of course, length on the waterline. Overhangs add to the efficiency of a given length on the L.W.L., and, when available, should always be adopted. But when from any causes the length overall is limited, then it is better from the point of view of economy to get as extended a waterline length as possible. In the case of very small yachts, say, of 18ft. to 24ft. overall length, where, as in the Humber Yawl Club boats, considerations of shipment have to be studied, the rule holds good, and though most of the boats of this club are sharp sterned they would be better cruisers, because actually larger and more convenient for the rig if they had transoms. This does not apply to the "canoe stern" boats, as here it is only a matter of choice between canoe stern and counter.

Having had a long experience in both transom sterned and sharp sterned boats, I think that even in running before a high sea the well-designed transom stern boat is better, and I have never seen any sea break over the stern of either a transom or sharp sterned boat. If beaching is a part of the boat's work, then it is incontestable that the sharp stern is the better form, but it must not be forgotten that in beach work the following wave has to be split by the stern of a vessel that is for the moment stationary. But cruising yachts are not taken to the beach voluntarily, and this particular quality is outside our consideration in dealing with a yacht's performance afloat.

The Design and Construction of Small Cruising Yachts.

BY

ALBERT STRANGE.

II.—UNDERLYING PRINCIPLES.

Displacement.

ALL floating bodies are governed by definite laws, one of which is that any unsubmerged form displaces exactly the same weight of water as its own weight. If you require proof of this fill a large jar exactly full of water and place it on a tray. Into this jar put an empty smaller one that will float. When it is put in, some of the water will run over out of the big jar into the tray. If you carefully weigh this spilled water you will find that it equals the weight of the small jar.

A totally submerged body does not necessarily displace its own weight. A cube of lead of 1 in. edge weighs .41 lbs. If this is put into a full jar of water it will only displace an amount of water equal to its *size, i.e.,* 1 cubic inch of water, which weighs .036 lbs. This is one of the reasons why lead is used as ballast; it is so much heavier than water per cubic inch or foot.

Therefore if, by calculation (to be explained later) the underwater body of a yacht measures 105 cubic feet, its displacement (or weight) is 3 tons exactly, because 35 cubic feet of salt water weigh one ton, and $35 \times 3 = 105$. It must be understood and remembered that the part of the yacht *above* the waterline is of no account in calculating displacement.

Do not confuse displacement with "tonnage." No one on earth knows what exactly "yacht tonnage" is, because it is a mythical measurement brought about by quite arbitrary rules. But "displacement" is a definite, clear, scientific term, and means the weight of the yacht with everything in and on her from the truck to the lead keel, whether she is afloat, or high and dry. Yacht designers think in displacements, not tonnages.

Centre of Gravity (C.G.)

All solids, and all plane figures, have a centre of gravity. It is the point on which they would balance if we could find it. A sheet of stout metal of uniform thickness is a solid—yet in one sense it is a plane. Its centre of gravity (CG) can easily be found if it is a square or a rectangle, by drawing lines from corner to corner. Where the lines cross each other is the point on which it will balance.

The centre of gravity of a yacht can be calculated to an approximate degree if the exact weights and position of timbers, ballast, furniture, mast, sails and equipment be known. But it is a lengthy and boring business which may be left until later, especially as its exact position has slight importance in the designing of ordinary shaped ballasted cruising yachts. In steamboats and motor yachts with large auxiliary sails it is a matter of greater moment, and its position should be ascertained.

The centre of gravity of a yacht or ship never moves from its position relative to the vessel, however the vessel may be placed —unless ballast or other weight is moved. Through it a force representing the whole weight of the vessel is constantly pressing vertically *downward*.

Centre of Buoyancy (C.B.)

This is a very necessary and important point to find, and it must be clearly understood as well as calculated. It is not easy to explain, but I will do my best. First of all, *it is not in the yacht,* though one of its positions is marked on the drawing. Secondly, *it moves about,* fore and aft and from side to side as the yacht moves in pitching or rolling. It used to be called the "Centre of Gravity of Displacement," and that is what it really is, only the term was found to be confusing, and it was altered by the scientific to its present more handy name. In short, it is a point through which a force equal to the weight of the displaced water is constantly pushing vertically *upwards*. When the yacht is at rest (in equilibrium) the C.G. pressing downwards and

makes in the water varies, although its volume is practically the same, and, of course, if the shape varies, the centre of gravity of the shape, *i.e.,* the centre of buoyancy, also varies.

Imagine a yacht pinned down by the head by a force (not a weight); suppose her anchor is immovably fixed in the ground and the tide is rising rapidly. What happens? The yacht gradually becomes depressed by the head, and so submerges her bow and lifts her stern. As she submerges her bow the shape of her underwater body forward alters and becomes much fuller. The C.B. shifts gradually forward to C.B.2, and as the yacht's weight, acting through

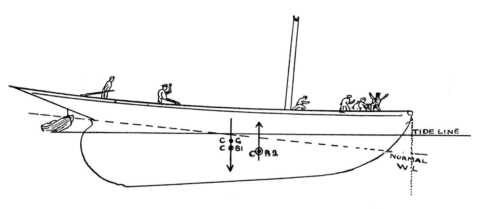

Fig. 7.

the C.B. pressing upwards act in the same vertical line, and thus neutralize each other. The force acting through both is exactly equal in all cases, because the weight of water displaced is equal to the weight of the yacht.

If we could solidify the water in which a yacht floats and then lift the vessel completely out, the hole in the solidified water would be exactly the shape of the yacht's bottom as she then floated. Now, if this hole were filled with, say, plaster of Paris, which, when it had hardened, was also lifted out we should have a solid model of the yacht's bottom. The centre of gravity of this model would be the position of the centre of buoyancy of the yacht in the upright position. But water is not a solid, and a yacht is not always in the same position. Therefore whenever the yacht heels, or pitches, the shape of the hole she

the C.G., remains in its position always pressing downward, and a force equal to the water displaced by the yacht is pushing upwards through C.B.2, it is easy to see that a mighty lever or couple will be formed and the C.B. and C.G., acting on its ends, will between them break the cable, and then the yacht will again float in her normal trim with the C.B. and C.G. once more in the same vertical line (Fig. 7). It will be seen that if the C.B. varies its position when the yacht is down by the head or stern, it will also vary when she is heeled sideways. The fact that it does vary whilst the C.G. remains stationary gives a vessel righting power when heeled under canvas.

Fig. 8 represents the section of a yacht heeled to port, and the C.B. has, in consequence of the altered shape in the water, shifted from C.B.1 to C.B.2. The upward pressure is again exerted through C.B.2, and the downward weight acts through C.G., creating a righting lever or couple, tending to bring the yacht to an upright position whenever the force that causes the yacht to heel (in this case the

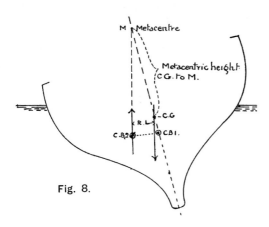

Fig. 8.

wind) lessens or is removed altogether. A vertical through C.B.2 will cut the inclined centre line in M. This is the Metacentre, which, as will be seen, varies slightly with different angles of heel. It is not an important point so far as keel ballasted yachts are concerned. But G must always be below M.

It will be noticed that the C.G. is above the C.B. in this diagram, which represents

all their ballast outside and very low down. It is not at all necessary that the C.B. should be above C.G.; in fact, a boat may be perfectly stiff and safe with it considerably below the centre of gravity; but, of course, the higher it is, and the lower the C.G. is, the better able is the yacht to stand up to her canvas, and the greater will be the Metracentric height shown on Fig. 8.

A very shallow hull of necessarily light displacement, such as a dinghy or a sailing canoe, though it may be ballasted, has its C.G. naturally very high above the C.B. After reaching a small angle from the vertical the two centres come into a vertical, and, of course, the least extra weight on the lee side will cause the formation of an upsetting couple, and the boat will inevitably capsize unless weight is promptly shifted to windward and the position of the C.G. thus altered and brought to windward of the C.B. Fig. 9.

The stability of most vessels of a cruising type is greatest at an angle of heel of from 30 deg. to 60 deg. from the vertical, provided the freeboard or height of side is ample and the displacement sufficiently heavy. When the deck becomes much immersed a great difference takes place in the shape of the plane of flotation, and the length of the righting lever gradually diminishes if the angle of heel goes on in-

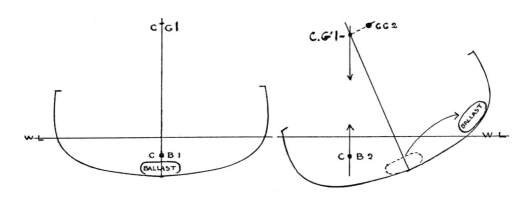

Fig. 9.

the approximate positions of these points in yachts with light metal keels or ballasted internally. The C.G. is rarely found below the C.B. in cruising yachts, unless they are lightly built and sparred, and have

creasing. Many yachts are perfectly safe, and have a good righting lever, even when laid down flat at 90 deg., but before allowing even the most stable of vessels to assume this position (which can only arise through gross carelessness, or else by what lawyers call " the act of God "), it is advisable to close all openings on deck, and secure the crew with strong life-lines. As a serious happening in the ordinary career of a cruising yacht it is almost outside the range of probability.

A vessel having a shallow hull and great beam, and consequently a very high position of C.G., should resist heeling very strongly, and thus be called very " stiff " at small inclinations, owing to the C.B. moving out to leeward very rapidly. But she will certainly ultimately capsize if the wind pressure is sufficient, and her danger angle will be reached at about 25 deg., according to the height of freeboard.

If, however, we fasten to this shallow hull a plate of metal or of wood with a lead weight at the bottom, we do two things : (1) we add to the weight (displacement) and (2) at the same time lower the centre of gravity of the whole yacht. If the boat were decked all over and quite watertight, she would now right herself, supposing she was laid over to an angle of even 90 deg., if the weight were sufficiently heavy and *sufficiently low.* (Fig. 10.)

What happens is that as the boat turns over from the upright to the position at 90 deg., the C.B. gradually shifts with the

relative to the boat, and thus produces a righting lever equal to the distances between the lines passing up and down through C.B2 and C.G. The forces upwards

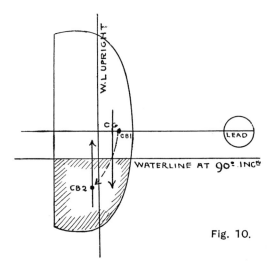

Fig. 10.

through C.B.2, and downwards through C.G., form a couple which will bring the boat slowly back to the upright, provided no water has got below, and that the crew (if any) are out of the boat. It will be noticed that in this extreme case the original position of C.G. is *above* C.B., proving that

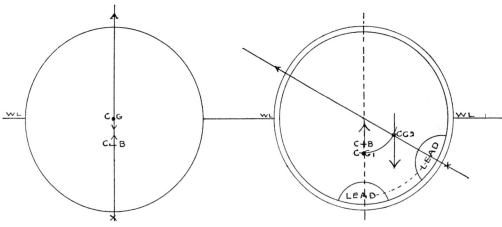

Fig. 11.

heel until it reaches the position marked C.B.2, the shaded part showing the shape immersed at 90 deg. equal in area to the immersed section when upright. The C.G., however, remains in its original position

stability can be secured under this apparent contradiction.

There is one more example I should like to give before leaving this somewhat dry, if extremely necessary, subject of stability.

We have hitherto dealt with forms of excessive and normal initial stability due to form. There is one shape, and one only, in which there is no change of form caused on inclination by heeling, and that is the cylinder. If we take a cylinder of light wood, such as pine or cedar, turned out of the solid, and if it is placed in the water it will float in a position similar to that shewn in Fig. 11. The C.G. will be in the centre, as shown, and the C.B. at C.B. If the cylinder is quite homogeneous throughout it will rest in any position it is placed in, shewing no tendency to return or depart from it, because the shape immersed always remains the same, and the C.B. is always exactly vertically under C.G.

Now, if we take this cylinder and hollow it out carefully, so that most of the wood is removed, it will, of course, float very much higher if the ends are closed. If a lead weight equal to the amount of wood removed be now placed inside, the C.G. will be lowered from C.G. to C.G.1, which represents its approximate position, and we change a completely unstable form to one of the most stable—one that has its greatest righting power when heeled to 90 deg., although the C.B. does not move its position (due to form) at any angle. The diagram shows how this happens. Although the C.B. keeps its place *relatively to the waterline,* the C.G. also keeps its place relatively to the position of the *added lead* in line AX, and by its small movement to C.G.2 sets up a couple with the C.B., which has its greatest effect when at right angles to it, but is here shown at a somewhat less angle.

The position of C.G.1 is, in this diagram, shown below C.B., but there would be a small amount of positive stability, even though it were above C.B. The mathematics of this pretty problem is to be found by those who care to work it out by " x's " on page 168 of Professor Attwood's learned little treatise on " Theoretical Naval Architecture."

The reader who has patiently followed this sketch of the elementary theory of stability, and thoroughly mastered its simple principles, need not go more deeply into the matter for the purpose of making a design. If he later learns to calculate the approximate position of the Centre of Gravity, he will know all that is usually put into practice by yacht designers, but only a tithe of that employed by the naval architects who design our warships and the vessels of our mercantile marine. The employment of lead or iron ballast in fixed positions, and the fact that a yacht never " changes trim " to a fraction of the extent that our merchant steamers do in carrying various cargoes or voyaging light with only water ballast, renders any deeper knowledge superfluous. But as the acquisition of knowledge, if only for its own sake, is a fine mental training; the student whose mathematical powers will allow him to study such works as Sir W. H. White's "Manual of Naval Architecture " and the books by Professor Attwood and other authorities, will, if he undertake the task, reap a rich reward.

The next article will deal with the simple calculations required in yacht design, and then we shall proceed to the actual planning of the yacht, having now well and truly laid our foundation of knowledge of the absolutely necessary underlying principles.

The Design and Construction of Small Cruising Yachts.

BY

ALBERT STRANGE.

III.—CALCULATIONS.

THE ordinary calculations necessary in preparing the design of a yacht are few and simple. They separate themselves into three divisions, as follows :

I. Areas of figures bounded by straight lines.

II. Areas of figures bounded by a straight line and a curve.

III. Volumes of solids of varying forms.

IV. The centres of gravity of all the foregoing.

The simplest figure bounded by straight lines is the triangle. The areas of triangles of any shape can be found by multiplying the vertical height by the length of the base and dividing the product by two. (Fig. 12.)

The areas of squares, rectangles and parallelograms are found by multiplying the base by the vertical height.

The centre of gravity of a triangle is found by bisecting any two sides, and from the points of bisection lines are drawn to the opposite angles. Where these lines cross each other is the Centre of Gravity (C.G.) (Fig. 14.)

The centres of gravity of a square, a rectangle or a parallelogram are found by drawing lines from corner to corner (diagonals). Where these lines cross each other is the C.G. (Fig. 15.)

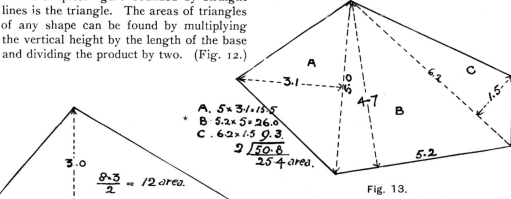

Fig. 12.

Fig. 13.

(*Note.*—The height is the vertical height from the base, *not* the length of side.) All other figures bounded by straight lines can be divided into triangles. The sum of the areas of these separate triangles equals the area of the given figure. (Fig. 13.)

To find the centre of gravity of a trapezium (the shape of a mainsail) divide the figure into two triangles. Find the C.G. of each (1 and 2), and join. Then divide the figure into two other triangles, and find the C.G. of these two new triangles (3 and 4). Join these again by a line. Where this line cuts the line joining the two first centres is the C.G.5 of the trapezoid. (Fig. 16.)

---

* Area B should be 5.2 x 4.7 and the total area of Fig. 13 therefore 24.6.
There are several instances where anyone with the benefit of a pocket calculator will produce a slightly different figure in the second decimal place from those produced by Strange, who, if he used any arithmetical aids at all, would only have had a slide rule.

To find the common centre of gravity of two different figures which are separate —such as a mainsail jib : First find the area of each, then find the C.G. of each.

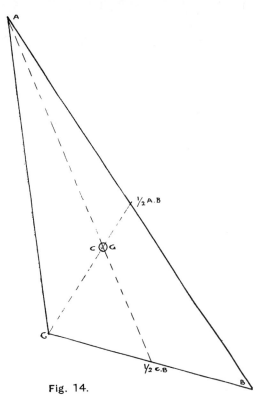

Fig. 14.

Join these two C.G. by a straight line. The common centre of gravity must be on this straight line. To find the exact position, *erect* a perpendicular at the C.G. of jib and *drop* a perpendicular at C.G. of mainsail. On the line erected from the C.G. of jib, cut off by any small scale a *distance* equal to the area of mainsail. On the perpendicular dropped from the C.G. of mainsail, cut off by the same small scale a distance equal to the area of the jib. Join the extremities, Y Z, of these two verticals. Where this line crosses that which joins the two centres of gravity of jib and mainsail is the point of common centre of gravity, or the "C.E." where sails are concerned. (Fig. 17.)

This method is also used in finding the C.L.R. of a centreboard boat when the board is down. First find the C.L.R. (or C.G.) of the lateral plane of boat, when the board is up in its case. Then find the C.G. of centreboard when down—that is, the part of the board exposed. Join these two centres, and from each centre draw verticals up and down. On these verticals set off by small scale distances equal to the areas of plane on board, *i.e.,* from the C.G. of board set off area of lateral plane of *boat,* and from the C.G. of lateral plane of boat set off by same scale a distance equal to area of *board.* Join the extremities of these lines, and where this line cuts the line joining the centres of board and lateral plane will be the C.L.R. of boat with plate *down.* (Fig. 18.)

To find the area of a segment, multiply the length of the base by two-thirds the vertical height, V. (This is approximate, but accurate enough for all practical purposes.) (Fig. 19.)

The C.G. of the segment will be somewhere in V. To determine its position, cut the shape exactly out of paper with V drawn in. Double the paper along V and balance it on a needle point. Where the figure balances is the C.G.

The position of C.G. of any regular-shaped figure can be found in the same way.

Another way to find the C.G. of a given figure is to cut the exact shape out of thin wood or thick cardboard and suspend the figure from two different points. A plumb line should be hung from each point of suspension, and a pencil line drawn across the figure on the line of the plumb. These lines will cross each other, and the point of crossing is the C.G. Care must be taken that the figure hangs freely and without friction from each point of suspension, or a large error is likely to happen.

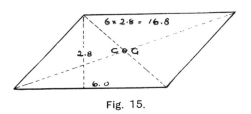

Fig. 15.

The calculation of the areas of vertical or "body" sections and horizontal or "waterline" sections: that is, of forms bounded by both straight and curved lines, are usually made in England by what is called "Simpson's First Rule.". Abroad it is quite usual to employ another rule, called the "Trapezoidal Rule," a much simpler

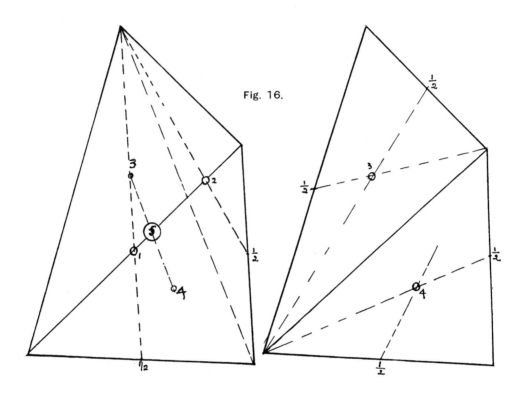

Fig. 16.

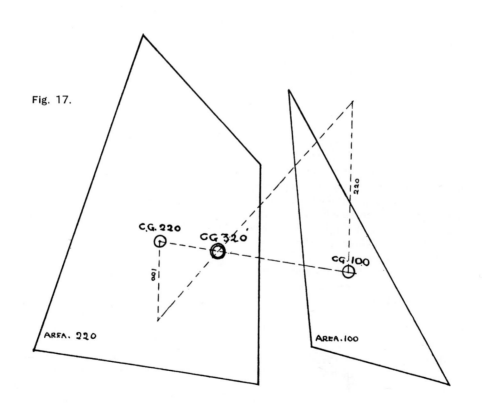

Fig. 17.

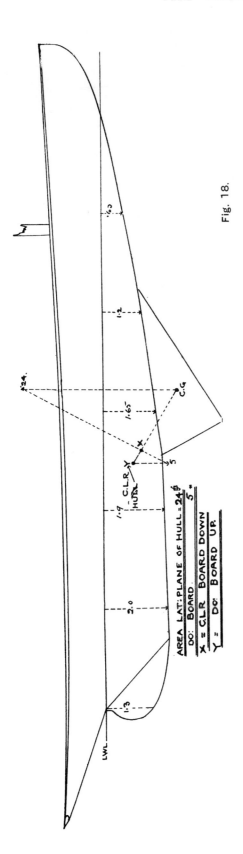

Fig. 18.

AREA LAT: PLANE OF HULL = 24 ☐
5 ·

DO: BOARD
X = CLR BOARD DOWN
Y = DO: BOARD UP.

affair, and one which presents several points of advantage for the amateur designer. For the areas bounded by strictly parabolic curves Simpson's First Rule is slightly more accurate, but for areas bounded by curves having contrary flexure —such as a 'midship section—the trapezoidal rule is as accurate. We will first take a figure bounded by a parabolic curve and compare the results of the two rules in calculating its area.

Simpson's First Rule is as follows : Divide the base line into any *even* number of equal parts (the greater the number the more accurate the result will be). At each division erect a perpendicular " ordinate " to touch the curve. The lengths of each of these ordinates must be measured to two places of decimals and noted.

As there are an even number of divisions there will be an odd number of ordinates. Number the ordinates 1, 2, 3, 4, etc.

Multiply the first and last ordinates by 1. The rest are multiplied in the following order : No. 2 by 4, No. 3 by 2, No. 4 by 4, and so on, until the end ordinate is reached. Add the products together, and multiply the sum by one-third the common interval (in this case 2 ft.) The result is the area of the figure. (Fig. 20.)

The working is as follows :

Ordinates		Lengths.		Multipliers.		Products.
1	=	0	×	1	=	0
2	=	.8	×	4	=	3.2
3	=	1.6	×	2	=	3.2
4	=	2.4	×	4	=	9.6
5	=	3.1	×	2	=	6.2
6	=	3.65	×	4	=	14.6
7	=	3.95	×	2	=	7.9
8	=	4.0	×	4	=	16.0
9	=	3.7	×	2	=	7.4
10	=	3.05	×	4	=	12.2
11	=	2.1	×	2	=	4.2
12	=	1.05	×	4	=	4.2
13	=	.1	×	1	=	.1

$$88.8 \times 2 \text{ ft. interval.}$$
$$2$$

$$\tfrac{1}{3})\overline{177.6}$$
$$\overline{59.2} = \text{area in sq. ft.}$$

The Trapezoidal Rule is far simpler. Divide the base line into *any* number of equal parts. At each division erect an

ordinate touching the curve. To *half* the
length of the first and last ordinate add all
the other ordinate lengths. Multiply by
the common interval, and the result is the
area of the figure. The convenience of
being able to divide the base line into any
number of equal parts is very great in
many cases. The area of the figure by the
Trapezoidal Rule is 58.9, and the area of
the figure as obtained by use of the Plani-

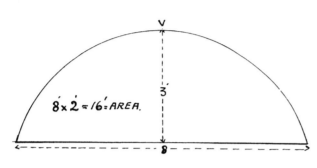

Fig. 19.

meter is 58.96 sq. ft. In these two examples the difference between the results is
not greater than the difference due to the
fact that only a closely approximate result
can be obtained by the use of two points
of decimals. There is a small error also
even with the use of the Planimeter, inas-
much as it is almost impossible to move
the pointer with absolute accuracy over the
whole figure. Where extreme accuracy is
desired when using the Planimeter it is
best to take three separate readings and
take their mean. But, as a matter of fact,
such extreme accuracy is only necessary in
very exceptional cases, such, for instance,
as in designing a racing sailing canoe or
model yacht, where "ounces" tell. The
greatest error that can occur when careful
use is made of either of the given rules
amounts to not more than one half per
cent. It is well to remember that with
parabolic curves the Trapezoidal Rule gives
an *under estimate* of the area by less than
one half per cent. This is not altogether a
disadvantage to the amateur, as the actual
area is a little more than his measurements
give, and is thus a margin on the right side.
The *over*-estimation or calculation of area
or displacement is far more troublesome and
disconcerting in its results.

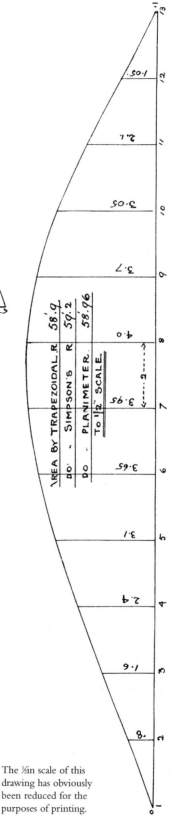

Fig. 20.

The ½in scale of this
drawing has obviously
been reduced for the
purposes of printing.

In measuring body sections there is hardly any difference between the two rules, as will be seen by the following calculation of a 'midship section. (Fig. 21):

Trapezoidal Rule (ordinates 1 ft. apart).

No. 1 = 7.0 ÷ 2 = 3.5   Area by Simpson's Rule = 30.70
„   2 =   = 6.75   Do. by Planimeter   = 30.75
„   3 =   = 6.20
„   4 =   = 5.28
„   5 =   = 3.80
„   6 =   = 2.20
„   7 =   = 1.40
„   8 =   = 1.10
„   9 = 1. ÷ 2 = .5
                ————
                30.73 sq. ft.

found are only useful for comparison with other designs, and are only exact when the yacht is upright and stationary.

It is impossible to locate the *exact* position of the C.B. when the yacht is moving and at the same time inclined, because the line of moving water surrounding the hull is extremely irregular, and never of the same shape for very long; but its position can be found for any *inclination* irrespective of movement, and as this is an important point to know it should be ascertained for at least one inclination. The most useful inclination to choose is when the yacht is inclined to nearly deck edge at lowest freeboard point, because then she may be as-

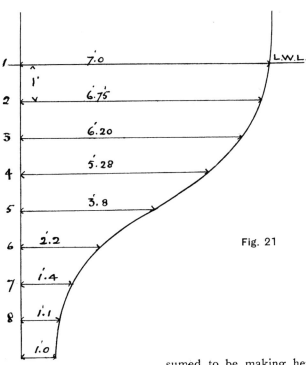

Fig. 21

Having mastered either of these rules, we are able to calculate the displacement of our yacht from the design, but there remain two other calculations which it is necessary to make, *viz.*, that for the longitudinal position of the Centre of Buoyancy, and the Centre of Lateral Resistance. Both of these are usually only calculated for the upright and stationary position of the yacht, but as both of these move when the yacht is inclined or in motion their positions when

sumed to be making her top speed. The method is as follows for the upright position: Take the half areas of the vertical sections and place them in a column. Multiply them by their positions in relation to No. 1. Add the areas, and divide the sum of the products obtained by the sum of the areas. Multiply the quotient by the distance the vertical sections are apart. As an example, we will take the areas of a yacht of seven tons displacement, 27 ft. W.L., whose sections are placed 3 ft. apart, the yacht being in an upright position. (Fig. 22):

Half areas in sq. ft.

Sec. 1 = 0.0 × 0 = 0.0   40.6)190.4(4.689
 ,, 2 = 1.2 × 1 = 1.2   1624   3
 ,, 3 = 3.7 × 2 = 7.4   ———   ———
 ,, 4 = 6.2 × 3 = 18.6   .2800   14.067
 ,, 5 = 7.5 × 4 = 30.0   2436
 ,, 6 = 8.0 × 5 = 40.0   ———
 ,, 7 = 7.0 × 6 = 42.0   .3640
 ,, 8 = 4.8 × 7 = 33.6   3248
 ,, 9 = 2.2 × 8 = 17.6   ———
 ,, 10 = 0.0 × 9 = 0.0   .3920
                             3654
        ———   ———
        40.6   190.4

14.067 is the distance the C.B. is aft of the fore end of L.W.L.

It will have been seen that the sum of the half-areas of sections amounts to 40.6. By the Trapezoidal Rule the displacement can easily be found from this, as follows :

        40.6 ft.
cu. ft.   3 = (distance sections are apart)
to ton ———
  35)121.8(3.48 = half displacement)
     105.   2 (to get whole displacement)
     ———   ———
     .168   6.96 tons.
     140
     ———
     .280
     280

This is a good example of the simplicity of the Trapezoidal Rule when used to find displacement. Not only is the working shorter, but in this case an even number of ordinates was taken to give a spacing of 3 ft. for the building moulds. This could not have been done if using Simpson's Rule. The weight and position of the Centre of Gravity of an iron or lead keel can be found in exactly the same way.

To find the Centre of Lateral Resistance the same method is adopted, using *lengths of ordinates* below water instead of areas (Fig. 22). The tabulation of the lengths will give the area of immersed longitudinal plane. The method is as follows, and the measurements are shown on Fig. 22, in which part of the rudder is included. The position of the C.L.R. in the actual yacht begins to move ahead as the yacht moves. The rounder the bow the further ahead it will go :

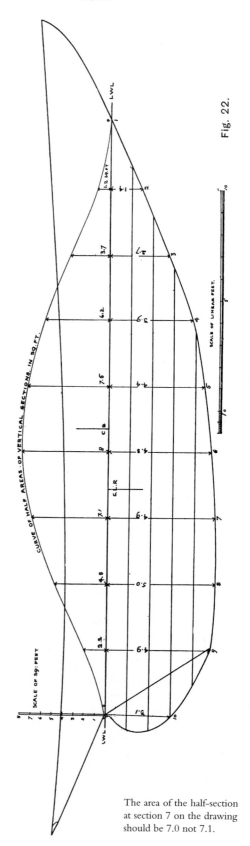

Fig. 22.

The area of the half-section at section 7 on the drawing should be 7.0 not 7.1.

Ordinates.	Lengths.	× Positions.	= Products.
1 = 0.0		× 0	= 0.0
2 = 1.4		× 1	= 1.4
3 = 2.7		× 2	= 5.4
4 = 3.9		× 3	= 11.7
5 = 4.4		× 4	= 17.6
6 = 4.8		× 5	= 24.0
7 = 4.9		× 6	= 29.4
8 = 5.0		× 7	= 35.0
9 = 4.9		× 8	= 39.2
10 = 3.1		× 9	= 27.9

35.1 ft. 191.6

$$\frac{191.6}{35.1} = 5.47 \times 3 = 16.41 \text{ ft. aft of [ord. 1}$$

To find area of lateral plane :

35.1
minus 1.55 (half of No. 10 ord., as required by Rule)

33.55
3 distance ordinates are apart.

100.65 = area of lateral plane.

The small piece aft of the last ordinate on the rudder, about 1½ sq. ft., can be added to the ascertained area. These are all the absolutely necessary calculations that the amateur need learn, at any rate to commence with. Several other different and minor ones will be treated of later, but if these are thoroughly mastered the task of commencing an actual design can be undertaken with confidence that every important element has been provided for.

It must be remembered that the "area of lateral plane" is only useful for comparison with that of other designs. It does not represent the area of wetted surface, which is obtained by another method to be explained later.

The Design and Construction of Small Cruising Yachts.

BY

ALBERT STRANGE.

IV. – [THE LINES PLAN STARTING THE DESIGN]

HAVING now gained a sufficient acquaintance with the principles of stability and the necessary calculations, without which we should be unable to ascertain the displacement, the different centres, or the sail area of any proposed yacht, we may now proceed to the actual operation of making the design.

A yacht design is a somewhat complicated problem in solid geometry, and to those who have never studied this branch of applied mathematics it will naturally, at first, present some difficulty in the making, though it is safe to say that many yachtsmen who could not make a design have a very good eye for reading the qualities of a yacht from her plans. Geometrical solids can only be represented on the flat surface of a piece of paper by the outlines of sections cut through the solids in various determined directions. One can easily see that any section of a sphere must be circular in outline, but the yacht's body varies in form in so many ways at all points of its surface that a large number of sections in various directions must be used before her lines or form can be read or built from.

The principal sections taken in making a yacht's design are these :

Vertical Sections *across* the hull from side to side, cut from the deck to the keel, called the Body Plan.

Vertical Sections, *along* the hull in a fore and aft direction parallel to the centre line, called Buttock Lines.

Horizontal Sections, *through* the hull, above and below the waterline and parallel to it, called the waterlines, and

Oblique Sections, *across* the hull, starting from the centre line, called diagonals.

All these sections cut one another at various points, which will be found to agree with each other in an accurate drawing. The difficulties arise in obtaining this exact agreement, which is only accomplished by the use of great accuracy of measurement, aided by good eyesight, for the difference between " fairness," or accuracy, and " unfairness " may be only small on one part of a plan, yet very large in another part. As both sides of a yacht should be alike in every respect, only one half need be drawn.

The centre fore and aft line and the centre vertical line are used as a base for this half. The fore and aft line is the base of the waterlines and deck outline shown on the half-breadth plan, and the centre vertical line is used for the base of the body plan. The fore body from the centre cross section to the stem is shown on the right side of this line, and the after body from the centre to the stern is placed on the left-hand side. But you can, if you like, place them just the other way, though there is no advantage in being unusual.

The only way to become familiar with these different sections is to patiently work them out from a very much simplified design, as shown in Fig. 24, where simplicity is at its simplest.

But before starting to make a design a few important details must be first con-

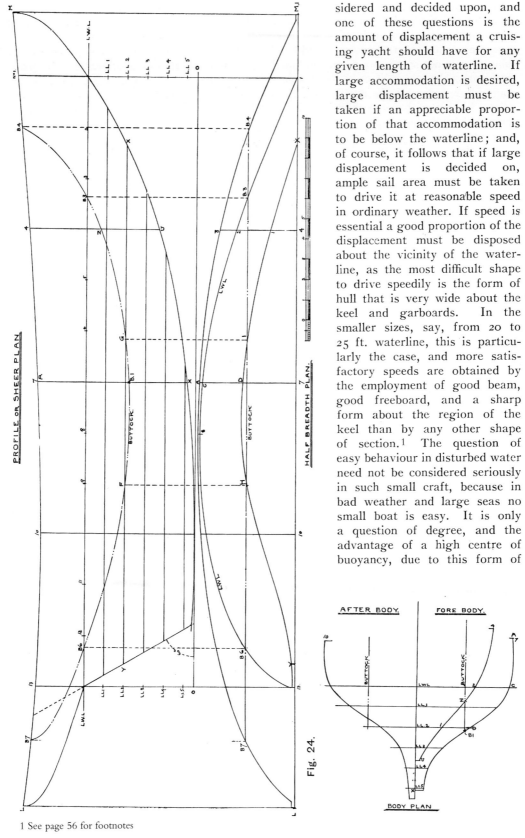

Fig. 24.

sidered and decided upon, and one of these questions is the amount of displacement a cruising yacht should have for any given length of waterline. If large accommodation is desired, large displacement must be taken if an appreciable proportion of that accommodation is to be below the waterline; and, of course, it follows that if large displacement is decided on, ample sail area must be taken to drive it at reasonable speed in ordinary weather. If speed is essential a good proportion of the displacement must be disposed about the vicinity of the waterline, as the most difficult shape to drive speedily is the form of hull that is very wide about the keel and garboards. In the smaller sizes, say, from 20 to 25 ft. waterline, this is particularly the case, and more satisfactory speeds are obtained by the employment of good beam, good freeboard, and a sharp form about the region of the keel than by any other shape of section.[1] The question of easy behaviour in disturbed water need not be considered seriously in such small craft, because in bad weather and large seas no small boat is easy. It is only a question of degree, and the advantage of a high centre of buoyancy, due to this form of

1 See page 56 for footnotes

section, must be taken as a compensating factor for any tendency to quick and uneasy motion inherent to this position of the most important centre of the yacht. Width of cabin floor is not so great a matter in small craft unless the owner or crew have abnormally large feet, and the ability to " keep on top " is better than the ability to plunge through a head sea by sheer weight and its momentum.

In order to assist the beginner in his estimate of the proper amount of displacement for any given length of waterline, the

The area of mid-section for 30 ft. W.L. is therefore at least 22 sq. ft. for the type of boat we are considering.

The proportions here given need not be taken as fixed and unalterable. In many cases the area of midsection may be increased, especially in the smaller sizes from 19 to 24 ft. W.L., in which the areas indicated on the curve may be taken to represent the minimum affording good room. As a whole the curve is fairly suitable for yachts of a proportion of from 3 to 3½ beams to length on W.L., a

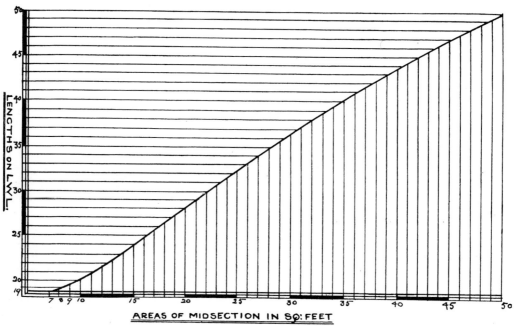

LENGTHS ON L.W.L.

AREAS OF MIDSECTION IN SQ: FEET

Fig. 23.

curve shown in Fig. 23 has been prepared from data obtained from many existing yachts of a cruising type, and represents the least average area of midsection apportioned to yachts of the waterline lengths given in the left-hand column. To use the curve, look up the left-hand column of W.L. lengths until the desired length is reached. Read along the ordinate from this length until the curve line is reached. Then the area of mid-section will be found by following the nearest descending ordinate from this point to the base line on which square feet are scaled. For example 30 ft. waterline touches the curve at the point erected from 22 sq. ft. on the base line.

draught of about ⅙ the length of W.L., and carrying a total sail area of the square of the waterline length.

An approximate estimate of the *displacement* (in cubic feet) is found by multiplying the length on W.L. by the area of midsection, and by .57. Thus, taking 30 ft. as a suitable length of W.L., the scale gives 30 × 22. To find displacement in tons the following formula is used :

$$\frac{30 \times 22 \times .57}{35.} = 10.74 \text{ tons.}$$

The divisor 35 is the number of cubic feet of salt water in one ton, and the constant .57 represents a prismatic co-efficient [2] a

2 See page 56 for footnotes

little fuller than the " wave-form " theory of a curve of versed sines for the areas of the fore-body and a " trochoid " curve of areas for the after-body would give.

If a somewhat fuller-ended model is desired, the co-efficient .6 may be substituted for .57. For the yacht of 30 ft. W.L. and 22 ft. M.S. this would work out as follows :

$$\frac{30 \times 22 \times .6}{35} = 11.31 \text{ tons.}$$

giving an increase of 5.2 per cent. in displacement. A larger co-efficient than .6 would give an unnecessarily full-ended boat, and if more displacement is wished for, it would be better to increase the area of midsection by a foot or two and keep the ends a little finer, or increase the length of L.W.L., keeping the same area of M.S. This latter would give a faster boat.

As most amateurs think out a yacht in terms of cabin accommodation rather than in terms of displacement in the first instance, the use of the curve of midsectional areas will be found very convenient in making the preliminary draught of the proposed yacht, as a sketch of the midsection would show floor room, head room and width of cabin.

There being no fixed method of procedure in designing, and as every designer seems to have his own private method, the following instructions must only be taken as *one* of the possible ways. It makes no claim to be highly scientific, and it leaves more latitude for individual fancy and experiment than some others. As yacht designing is an art as well as a science, this is as it should be. At any rate, the method will, in the hands of one who has a good eye for a fair line, produce a fair and shapely boat of good displacement and sufficient accommodation for usual requirements. Special requirements must be dealt with in special ways, but even for these the method will be found to give satisfactory results.

We will choose for our first attempt a vessel of moderate size—30 ft. W.L., 10 ft. beam, 5.5 draught, with a displacement of about 11 tons. By the scale we have already ascertained the area of midsection —about 23 sq. ft.[3]—giving approximately 11.3 tons. This will be ample, and if considered too great, a shade less area of mid-

section can be taken, though it is best to have a trifle of displacement in hand.

The best scale to use for a boat of this length will be 1 in. to the foot, but ¾ in. is a very handy scale. A scale of half an inch to the foot is a little too small for this size of yacht, and if adopted for sheer and half breadth plans the body plan should be enlarged to 1 in. scale for calculations. Whatever scale is used must be decimally divided, *i.e.*, the feet must be divided into tenths, not inches.

First sketch your midsection and get the area correct (only one half need be drawn). Draw the ordinates L.W.L., 1, 2, 3, etc., one foot apart by scale. These are the waterline ordinates for this section. Use the trapezoidal rule to find the area. (Fig. 21, Article III.)

Having decided on the shape of our midsection in the sketch, transfer it to the large sheet of paper on which the design is to be made. Place the centre line for midsection as far to the right hand side as it will conveniently go, and about a foot or fifteen inches from the bottom edge of the paper. Carry the line marked L.W.L. right across the paper to the left, and somewhere along this line mark off 30 ft. by scale, taking care to leave plenty of room clear of the midsection for the fore overhang and also for the counter at the other end. (Fig. 24.) [4]

In deciding how many sections we will use, thought must be given to the method of construction. In this case the grown frames may be placed 30 ins. apart centre to centre. It must also be remembered that the more divisions we make the more accurate will our calculations prove. The waterline is often divided into ten equal parts by many designers, but twelve is a better number, and as we wish to be accurate and save labour in the construction drawing, we will take this number of divisions, as the Trapezoidal rule allows any number to be used. So divide the 30 ft. length into 12 accurately equal parts, and giving 2 ft. 6 in. space between centres; but at first only draw verticals through 1, 4, 7, 10 and 13, as these are the only points that we shall use in determining our preliminary sections. These verticals must be produced in lightly drawn pencil lines above and below the L.W.L., in order to enable us to draw the preliminary sketch lines for the profile, which we will now proceed to do. Through A, on midsection on profile sweep in a line for the edge of deck, letting it rise in a gentle curve until it passes through a point on No. 1 about 3 ft. 6 in. or so above L.W.L. Aft of A this sheer line should descend slightly until it passes through a point on No. 10, about 2 ft. 3 in. above L.W.L., this being the point of lowest freeboard. Aft of this the sheer line curves gently up. Do not determine the length of sheer line yet; let it sweep away indefinitely at each end until the rest of the profile is determined.

The height of freeboard chosen is ample, and allows for an angle of 25 degrees heel before the deck, at least freeboard becomes immersed.

Now proceed to draw the underwater part of the profile. At a distance of 5 ft. 6 in. below the L.W.L. draw a long horizontal line—O. This is to represent the building level, and it fixes the chosen draught of water. At the point where No. 10 touches this line commence to sweep in the outline of keel. This should be a gentle curve passing up through X, then rising somewhat more sharply with a round convex curve, passing through the lower part of line No. 4 and up through the point made by No. 1 on the L.W.L. Continue this curve more sharply still, letting it rise at the end with a sharp, almost vertical, rise until it meets the sheer line. The shape of this curve should receive careful attention, and may call for several corrections before it is satisfactory. Do not attempt to pin it in to any definite length yet.

At 13, on L.W.L., draw the sternpost at an angle of 60 degrees with L.W.L. This amount of rake is sufficient for handiness and sufficient to ensure steadiness in running. It should not be exceeded in a cruising yacht.

Before leaving the profile for the half-breadth plan it is necessary to remark that a sheer line always looks more pronounced in the drawing than it does in the actual yacht. This is due to the fact that the sheer line not only curves *upwards,* but also *away* from the eye at each end. The longer the overhangs the less sheer needed for appearance. The lowest part of the sheer line should never be ahead of about one quarter of the length from the aft end of L.W.L.

To commence the half breadth plan draw L.M. a little over 5 ft. by scale below the line O, and from the profile square down lines from the extremity of stem and Nos. 1, 4, 7, 10, 13, and end of counter. At 7 set off from the midsection in body plan the breadth on deck and at the water line at A and C. Draw a preliminary deck line, passing through A and ending at M1, and L1 and see that it is round enough aft. We have now one point, C, through which the load waterline must pass, and the point at each end where it must begin and terminate.

The actual shape of the load waterline cannot be determined by any formula. It must be drawn by eye, and should be a full round curve in this particular case, without hollow forward. Its greatest width must be about the region of 8, 2 ft. 6 in. or more aft of the midship section, and its after part must be full and round also, if, as in this instance, a counter or canoe stern is to be adopted. If it is very fine aft, the buttock lines will not be convex, but hollow, unless the deck is unduly pinched in and the overall length curtailed. A very flexible batten must be used at the after end to get the required round, or else a French curve must be employed. The flexible batten is a far more satisfactory implement, however carefully the curve is used.

Now we must, before going further, get in the mid-buttock line. As already stated, the buttock represents the lower outline of a vertical fore and aft section, taken through the yacht in a line parallel to the centre line, but at some distance out from it; in this case, 2 ft. 6 in. representing the quarter beam. So, from L.M. measure 2 ft. 6 in. up 7, and draw a horizontal line parallel to L.M. at this distance. This line crosses the L.W.L. at B3 and B6 on half breadth plan. Square up these points to the corresponding letters on the profile, and also square up B4 and B7 to touch the sheer line on profile.

Turn to the body plan, and each side of the centre line draw verticals 2 ft. 6 in. out each side (these lines are marked buttocks). Note where the one on the right cuts the 'midship section at B1. Measure the distance to B1 from the L.W.L. on body plan, and transfer this distance to B1 on 7 on the profile. We have now five points in the curve of the buttock line, B4, B3, B1, B6, B7. With a light flexible batten pinned over the middle three of these spots draw a fair round curve through B6, B1 and B3, sketching the steeper curves at the fore and aft ends by hand, and with a light line, for these are the difficult places in the buttock line, and it is unlikely that they will be correctly drawn first of all; they will certainly require subsequent correction. The curve of the buttock line below L.W.L. should be of a parabolic nature. It looks well on paper if it follows, to some extent, the outline of the keel forward, but it is by no means necessary or essential that it should do so exactly.

It is now necessary to get in another waterline or horizontal section, so draw LL2 across profile, square its end down to X at stem, and Y at sternpost. These points are where it begins and ends in half breadth plan. Take the distance LL2 to D on 'midship section, and set it up at D on 7 on half breadth plan. We have now three points through which to draw, but by dropping lines from the points F and G on profile when the buttock crosses LL2 we get two more, H and I, on the line marked Buttock on half breadth plan. Through these points X, I, D, H, Y, draw a curve for LL2 outline, letting it be a similar curve to that of the L.W.L. forward, but having a hollow aft with a quick turn at its after extremity.

We have now some guide towards obtaining the shapes of sections 4 and 10, and will proceed to put them in. On the right hand side of centre line on body plan cut off 4, 2 on L.W.L. and 4.1. on LL2, taking these distances from half breadth plan. Set off also 4.3 at its right height from L.W.L., and the depth from L.W.L. to underside of keel at U, taking the distances from profile. Sketch the curve of section 4 by freehand through these points, and we shall have approximately the shape of this section. Check the shape by measuring the distance from L.W.L. to where the curve of buttock crosses 4 on profile, and measure this distance down buttock on body plan at Z. This gives another point on the curve of section 4 on body plan.

The same procedure is followed throughout on section 10, which should be worked on the left-hand side of the centre line of body plan. This section will have a good deal of hollow in it, but the curve must be fair and regular. Calculate the areas of each section as you proceed.

Having followed these directions and drawn line for line in the order given, there should be no difficulty in completing the first sketch of the design, and putting in all the remaining sections and waterlines. All the horizontal and vertical lines may now be inked in. Surprising things will happen at first, especially at the after end of the yacht, but a scale is given, and, if carefully used when copying the skeleton design, these surprises will not be so alarming. It is better for the beginner to copy two or three designs first, before proceeding to work on an original idea. This will save a lot of time and trouble, and disappointment will be avoided. As experience is gained, difficulties will be foreseen and anticipated. In completing the design proceed by waterlines, letting these follow the outlines of those already in place. If you attempt to draw the vertical sections first it will be more troublesome to get them fair. The battens will help to keep the form easy and flowing, and the curves of waterlines are far easier to judge than the curves of the sections with their sharp turns can possibly be in the early stages of a design. The very thorough articles on the use of the battens, weights and instruments written by M.I.N.A., and published in THE YACHTING MONTHLY, in Nos. 81 et sequitim, 1912-13, render it unnecessary for me to enlarge on this part of the subject, as it is presumed that the amateur designer will be conversant with the methods of using them, even if he is not a master of all their difficulties and contrariness.

In completing the skeleton design the chief places where trouble may be expected are forward and aft above the waterline. A series of level lines above the L.W.L. should therefore be put in and very carefully checked at their intersection with the buttock lines. These are shown in the completed design by LLA and LLB. (Fig. 25.)

The design is further " faired " by means of diagonals. These are a set of oblique sections put in across the body plan and developed on the blank half of the half-breadth plan on verticals set off at the sections stations. In the completed design they are set off from L.L.B. instead, as the curve of sectional areas is placed on the lower side of the half-breadth plan. It does not matter where they go on the design.

It will be plainly apparent how the diagonals are developed. The distances from the centre line of body plan to the extremities of each set of diagonals are set off on the vertical lines of the sections they end on and a curve drawn through the points. This curve must be continuously fair in every case, and if the batten will not touch all the spots set off, the body plan must be amended until it does. An alteration to one spot may mean an alteration to several others. It is a long business at first, but it is essential that it be completed accurately. At least three diagonals are required.

The curve of sectional areas is perhaps more difficult for the beginner to understand, but if he will now consider *areas* and not *lines* it may become clear. At each cross section *the area of that section* is set off by some small *linear* scale, and the ends of the *lengths* representing the *areas* connected by a curved line, which must be fair. If the body of the boat is fair, these areas will come fair, though not necessarily in a pretty curve. In this instance the areas are marked on the ordinates, and a scale of one-fifth of an inch was used. If too large a scale is taken the curve develops a big hump, and is difficult to draw. The *meaning* of the curve is to show in what order the water is displaced, supposing the vessel moves along in an upright position, and to the practised eye it gives a good idea of the fore and aft fineness of a vessel. It was invented by the great naval architect, Colin Archer, of Laurvig, and immediately superseded the rather fanciful " Wave Line Theory " propounded by Scott Russell. It is in universal use for all classes of vessels, and can be used with any form of waterline. There is no doubt that it represents an easily driven hull.

Archer's theory was that the fore body of a vessel, from the centre of buoyancy to the stem, should have its successive areas of sections arranged so that the curve containing them should take the form of a "curve of versed sines." The areas of the sections of the body aft of the C.B. should be bounded by a curve called the Trochoid. In the completed design this rule has been adhered to absolutely strictly aft, but with the usual allowance of five or six per cent. off the fine part of the forward end. It is well, however, to mention that some designers prefer a slightly finer curve aft, something between the shape of the Trochoid and the Curve of Versed Sines. At any rate it should never be fuller than a

Trochoid aft, and never *quite* as fine as a Curve of V. Sines forward, or the C.B. will be unduly far aft of the centre of L.W.L.[5]

The design given here is not set forth as being incapable of improvement, but it may perhaps be stated that it has actually been built from, and the yacht is reported to be a successful cruiser, fast in moderate weather and a good sea boat. [6]

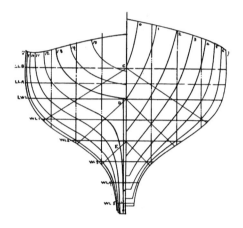

CALCULATIONS.

Herewith are the necessary calculations by Trapezoidal Rule :

Displacement.			Centre of Buoyancy.	
Sections.	Half areas in sq. ft.		Multipliers.	Moments.
1 =	0	... ...	× 0 =	0
2 =	1.0	... ...	× 1 =	1.0
3 =	2.94	... ...,	× 2 =	5.88
4 =	5.66	... ...	× 3 =	16.98
5 =	8.0	... ...	× 4 =	32.0
6 =	10.2	... ...	× 5 =	51.0
7 =	11.5	... ...	× 6 =	69.0
8 =	11.7	... ...	× 7 =	81.9
9 =	10.8	... ...	× 8 =	86.4
10 =	9.2	... ...	× 9 =	82.8
11 =	6.3	... ...	× 10 =	63.0
12 =	3.0	... ...	× 11 =	33.0
13 =	0	... ...	× 12 =	0

(*Note* that the multipliers begin with 0, not 1)

Total   80.3 × 2.5 (distance apart   522.96
       2.5   of section)   ───── = 6.5 × 2.5
cub.   ─────   80.3
ft. in 35) 200.75 (5.73 = half displacement
ton   175   2

.257   11.46 = tons (whole displacement).
245   6.5
─────
.125   2.5 = distance of sections apart
105   ─────
         16.25 = dist. of C.B. aft of No. 1

See page 56 for footnotes 5 & 6

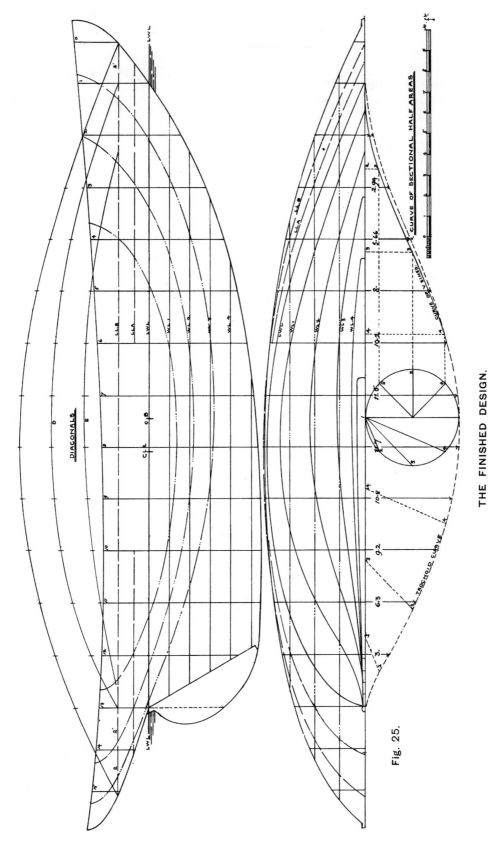

Fig. 25.

THE FINISHED DESIGN.

Length, over all, 39 ft. ; Length, on waterline, 30 ft. ; Beam, at deck, 10 ft. ; Beam, extreme, 10.1 ft. ; Draught, extreme, 5.5 ft. ; Displacement, 11.46 tons.

One other calculation has now to be made, and that is to find the position of the Centre of Lateral Resistance, in order to enable us to judge where the Centre of Effort of Sails should be. This calculation is exactly similar in working to that of the calculation of the C.B., only the *depths* below the W.L. of each section are used, instead of areas.

CENTRE OF LATERAL RESISTANCE.

Sections.	Depths.	Multis.	Moments.
No. 1 =	0.0 ×	0 =	0.0
,, 2 =	1.55 ×	1 =	1.55
,, 3 =	2.8 ×	2 =	5.6
,, 4 =	3.7 ×	3 =	11.1
,, 5 =	4.3 ×	4 =	17.2
,, 6 =	4.8 ×	5 =	24.0
,, 7 =	5.1 ×	6 =	30.6
,, 8 =	5.35 ×	7 =	37.45
,, 9 =	5.5 ×	8 =	44.0
,, 10 =	5.5 ×	9 =	49.5
,, 11 =	5.5 ×	10 =	55.0
,, 12 =	5.3 ×	11 =	58.3
,, 13 =	3.7 ×	12 =	44.4

Sum of Depths = 53.1   378.70 = sum of Moments.

$$\frac{378.7}{53.1} = 7.11$$

7.11 × 2.5 = (distance of sections) = 17.775 ft.

17.775 ft. is the distance the C.L.R. is aft of Sec. 1.

The calculation of displacement by Simpson's Rule gives 11.52 tons, and is probably nearly exact. But the difference between the two is only 1.4 cwt. in 11½ tons. [7]

There are several ways of simplifying the calculations by means of arithmetical contractions, but I have thought it best to give results in full, and in the simplest manner, so that the processes can be followed by all. It will be observed that a part of the rudder is included in the calculation for Centre of Lateral Resistance.

The area of the largest section, No. 8, is 23.4 sq. ft. This amount treated by the formula for estimating approximate displacement, *viz.,*

$$\frac{23.4 \times 30 \times .57}{35}$$

gives 11.42 tons, a very near approximation.

---

1 The lightening of construction since the advent of GRP and modern composites has allowed this principle to be taken further than Strange could have imagined.

2 Prismatic Coefficient: A comparison of the volume of the underwater body with that of a block (prism) whose length is the waterline length, and whose constant cross-sectional area is equal to the area of the immersed midsection. This ratio is also the ratio of the curve of transverse areas (see page 54) to the area of the rectangle which circumscribes the curve of areas.

3 For 23 sq. ft substitute 22.

4 Space has not allowed the printer to show the midsection drawing in the position described by Strange. An example of a lines plan where it is so positioned can be found on page 124.

5 To construct the Curve of Versed Sines and the Trochoid. (Referring to Fig. 25, p.55).

Mark a point on the fore and aft centre line of the half-breadth plan exactly below the Centre of Buoyancy. Join this point to the Curve of Areas by a line parallel to the section lines. With this line as diameter, scribe a circle. Divide each semi-circumference into four equal parts. Similarly divide both parts of the fore and aft centre line (total length equal to the LWL) into four equal parts, and draw perpendiculars down from the fore part of the line.

The Curve of Versed Sines. From points 2, 3 and 4 on the fore semi-circumference, project lines parallel to the centre line. The points at which the ends of these lines meet the perpendiculars previously drawn plus the fore end of the LWL and the lower tangent point of the diameter, define five locations on the required curve.

The Trochoid. Join points 2, 3 and 4 on the after semi-circumference, to the diameter's upper tangent point. Parallel with, and of the same length as, these chords 1-2, 1-3, 1-4, draw lines 2-2, 3-3, 4-4 on the after part of the fore and aft centre line. The ends of these lines plus the after end of the LWL and the lower tangent point of the diameter, define the Trochoid.

It is worth noting that although Archer's system of using these curves to establish the distribution of the immersed volume of a hull is no longer in use, it nevertheless produced satisfactory results.

6 This statement presents a conundrum. *Cloud* (see page 127f.) is the obvious contender, yet the differences are just too many to be ignored. No record of the building of another yacht of this size at this time has been found. It is possible that Strange re-drew the design of *Cloud* for the purposes of this series, with minor adjustments suggested by experience, or merely without the constraints imposed by a client.

7 Strange is here comparing the displacement calculated by the Trapezoidal Rule (on page 54) as 11.46 tons, with 11.52 tons by Simpson's Rule. In the following paragraph, in which he gives the displacement by the 'quick' formula, he has reverted to a value of .57 for the prismatic coefficient, where earlier (see page 51) he was using .6. It seems that he took his own advice (in the second paragraph of that page), and increased the midsection from the 22 sq. ft, suggested by his graph in Fig. 23, to 23.4 sq. ft rather than adopt the larger coefficient of .6.

The Design and Construction of Small Cruising Yachts.

BY

ALBERT STRANGE.

V. – [A DISCUSSION OF SHOAL-DRAUGHT CRAFT – SOME LOCAL TYPES]

THE design given as an example of a cruiser of ordinary type in our last chapter was used as an illustration of the method of commencing the work and making the calculations necessary for its completion. It presented no new problems or any difficulties of form, and may be regarded simply as one of many hundreds of very similar yachts in actual existence all round our coasts.

When one considers the characteristics of a large part of our eastern sailing waters it becomes evident that a yacht of even such a moderate draught would not always be the best type for general cruising. It may truly be said that apart from the deep water available in the south and south-west and on the Scottish coast, the greater part of our seaboard is of such a nature as to compel the use of yachts of very light draught. Such craft call for very special knowledge on the part of the designer, both in form and construction, which can only be adequately gained by actual acquaintance with local conditions, and experience of the behaviour of such yachts under a wide range of circumstance. It was, in fact, the lack of suitable cruising craft for shoal waters that many years ago first led me to the study of yacht designing, and it may be that many men are now in exactly the same position of being unable to obtain the most suitable craft for their requirements unless they are designed by themselves; and it is mainly for such as these that these articles are written.

It is a curious fact that in England hardly any attention has been specially devoted to the design of light draught or centre-board cruisers, and very few designs of such have been published outside the year books of one or two local clubs, of which the Humber Yawl Club is perhaps the best known. It is true that light displacement centre-board boats are fairly numerous, but they are little more than day boats. The outcome on the Thames of the search for a shallow draught cruising boat has been the production of the barge yacht by Mr. E. B. Tredwen, who has for years past devoted much energy and resource to its improvement, and who can fairly claim to have brought the type to its present state of usefulness and efficiency.

Certainly the barge yacht does, for its length, draw the least water of any existing shoal draught type of yacht in which it is possible to cruise, and in this respect cannot be improved upon. If it has not become so multitudinous as one might expect, it is, I think, due to the fact that, while admitting its ability to sail where other types would be hard and fast aground, most men are willing to sacrifice something in the matter of draught on the altar of appearance, and, moreover, a knowledge (almost instinctive) that at some angles of heel recovery is really impossible, has deterred others from possession or from building. That they are safer than many suppose, when handled by experienced men, is indisputable in view of the published accounts of their voyaging, but it will be obvious to those who have digested the previous article dealing with the question of stability that the last objection has its foundation in fact.

In America, however, the shoal draught cruiser is ubiquitous, and has reached a very high stage of development. Not only

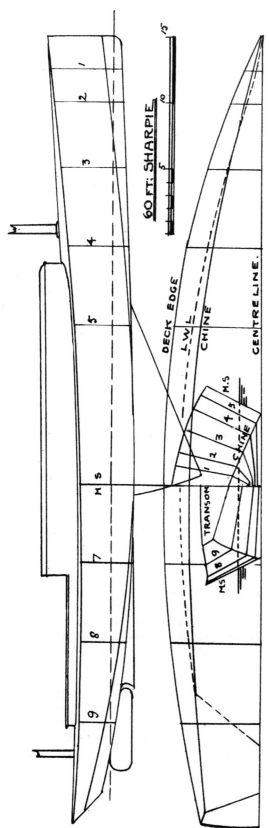

60 FT. SHARPIE.

is the hard-bilged Catboat extensively used on the eastern seaboard, sometimes doing arduous work in exposed channels in almost every kind of weather; but the other type, with a square, angular bilge, called sharpies, not dissimilar in principle from the barge yacht, but having little fore and aft " flat " and sharper ends, is extremely popular on the more sheltered bays and inlets. Until the motor finally drives all sailing craft off the face of the waters they will continue to exist, if not to increase, in numbers. (Fig. 26.)

But apart from relatively small craft having extremely light draught, quite large yachts and trading vessels are still constructed and used having the same useful characteristic. All these different types have one thing in common—they are all fitted with a centreboard—and to the seagoing man of British birth this is their chief objection. Where the American will use the centreboard for trading or pleasure vessels, the Anglo-Saxon will use the leeboard in preference. The Dutchman and the Belgian follow the same path, partly from tradition, partly because in a few points the leeboard is constructionally simpler, but mainly because the centreboard is not completely understood, and is apt to be troublesome if carelessly used, or if the necessary strength of construction has not been secured. There has been, too, a mistaken idea in using centreboards of excessive weight requiring machinery to hoist and lower them, and also the error of placing a centreboard in a yacht which already possessed ample lateral plane has not been infrequent. The most outstanding example in this respect was the Queen Mab, a yacht of 59 ft. on the W.L., 16 ft. beam, and draught of 11 ft. This yacht, after winning many races when fitted with a heavy metal centreplate, was found to be quite as fast when it was removed. Its absence did not affect her victorious career, and, until she was sold to America, she remained the fastest of our 40-raters. The proportions of this celebrated yacht, it will be seen, approach very nearly those that have been given in a previous article as being suitable for a cruiser of approximate length, and they were almost the first to be adopted in the racing classes. She paved the way for the retreat from the position that had been held so long by English designers, that beam was detrimental to speed and seaworthiness.

The idea that a heavy plate adds greatly to the stability of a yacht when it is lowered is greatly exaggerated. If a plate weighing six hundredweight is placed in a yacht of four tons displacement, and given a total drop of 4 ft., the centre of gravity of the whole yacht is only lowered .3 ft. when it is down, as may be seen by the formula :

$$\frac{Wt \times Distance}{Displacement.} = \frac{.3 \ tons \times 4 \ ft.}{4} = \frac{1.2}{4} = .3$$

Such a plate is unusually heavy for a yacht of the size, and such a drop is much greater than is usual, but it will be seen that even in a case where all the advantages possible are granted the gain is not so enormous as one might imagine, and it is not worth the cost and danger of straining.

In America, where the centreboard has been well studied, large boards are rarely made of any other material than wood, and a board of 12 ft. exposed area can be lifted easily by the simplest of tackle. Such a board is ample for a yacht of 30 ft. waterline, and although the case does take up a considerable amount of cabin space, it must not be forgotten that the large beam which is not only allowable, but positively necessary in boats of a shoal type of body, mitigates this nuisance.

Leeboards, on the other hand, do not take up cabin space, but are not so effective in holding a yacht from driving to leeward, for two reasons. Their area is generally smaller, and usually the angle they make with the water is not an effective one unless the yacht is actually designed to prevent this ineffective angle, by having, as is seen in the Dutch types, an exaggerated amount of tumblehome in the topsides. They are unsightly from the sailor's point of view, though the artist eye approves their picturesqueness; they are noisy at anchor, and they require attention on every tack unless working through guides.

The plan adopted in some racing yachts of Colonial type of having a plate fitted in each bilge has much to recommend it in small craft. But in large and heavy yachts the cutting away of so many timbers each side is fatal to strength and tightness. This alone should put them beyond consideration so far as cruising qualities are concerned.

We are brought, then, to consider the relative advantages of the centreboard type and the leeboard type in our search for the best kind of hull for use in shallow waters and dry harbours. The root idea should not, I think, be the question of "how little," but the exact opposite—"how much" draught can be adopted to ensure not only a big range of action, but as large an amount of actual seaworthiness as possible. The fact that a yacht can sit absolutely upright on the ground is doubtless a great asset to those who desire to spend a large part of their cruising time in a stationary position. It is most convenient where the cruising ground dries out at half ebb for many miles, and it is a fact that a flat-bottomed yacht, such as a barge or sharpie, does not pound on the sand when being floated by an incoming tide, the water under the yacht forming a cushion until she is afloat.

Against these advantages must be placed the fact that all experience has proved over and over again that the better performer a yacht is on the ground, the worse she behaves on the sea. The more nearly she will sit upright, the more she will drive to leeward in a breeze and a beam or head sea.

There is one exception to this statement. One type does exist that is of shoal draught and is yet, with skilful management, a good, if not exactly safe seaboat. This is the coble of the N.E. coast, a marvellous example of man's indomitable ingenuity in overcoming seemingly insuperable obstacles. The example given as an illustration was measured from an exquisitely built model in my possession, dating certainly from 1850, perhaps earlier. It displays accurately all the characteristics of the type, whose origin is undiscoverable, though doubtless it has, like all other types, been produced gradually, countless experiments having refined and perfected its extraordinary virtues as a working fishing boat. The only differences discernible in the coble during the last sixty-five years are a slight increase of draught forward, and the rounding in of the stern to a sharp point. Boats having the sharp stern are called mules, and are said to run better in a sea. Running is the weak point of all the boats, and it only requires a glance to see that with the deep forefoot this will be the case. The rudder is, of course, unshipped

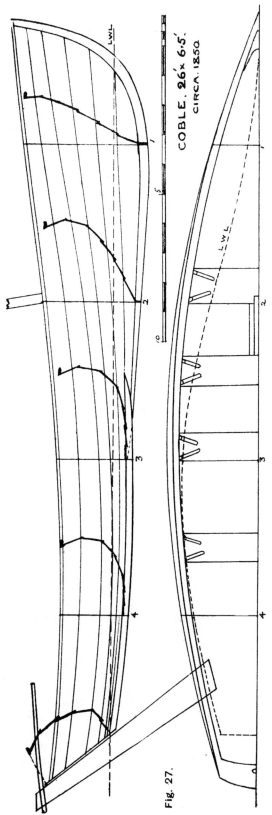

COBLE. 26' x 6.5'.
CIRCA. 1850.

Fig. 27.

when the boats are brought to the beach. Should a rudder break, a fatal accident is almost inevitable, particularly in running. On a wind, especially in coming to the land from the fishing grounds four or five miles to leeward, ballast is shifted to windward, and one of the crew of three is constantly baling, the boat being sailed hard under her single lug, which is never dipped on short tacks.

The sketch is given as an illustration of a unique type of light draught sailing boat, *not* as a model for a cruiser. A more unsuitable hull for any purpose of cruising could not be devised. (Fig. 27.)

A larger size of boat, from 30 ft. to 33 ft. over all, with greater proportional beam, drawing from 3 ft. 3 in. forward to a foot or so aft, is used for line fishing and for herring drifting. Boats of this size have been built for cruising yachts, but the rudder difficulty is almost insuperable.

The fine hollow bow, moderate beam and flat floor; in fact, the whole underwater form, is identical with that of the modern high speed motor boat hull, a striking illustration of the anticipation of the conclusions of science by the unaided exercise of skill and deduction on the part of uncultured fishermen.

There is practically no difference in hull form between the coble and the sharpie under water, which seems to indicate that for very flat-floored craft of this kind moderate beam, sharp lines forward, and a clean run are indispensable for speed. At the same time the structural advantage lies with the coble, both in strength and lightness, and although the type is stiff at first and through a small range of heel, its stability soon vanishes. A coble is never sailed with the sheet fast, and a sharpie requires just as much care in handling and an equally small amount of sail area in proportion to overall length.

Leaving the flat-floored angular bilged craft therefore to those whose requirements compel their use, let us consider the more yacht-like forms of centreboard boats suitable for cruising in shoal waters, such as those of our numerous estuaries.

The catboat pure and simple, with its draught limited to about one-tenth of the waterline, and having but one large sail set on a mast placed absolutely as far forward as it can be, and a very large wooden board, has never been popular in English waters. It is a fast type, especially to windward, so long as there is no sea to contend with. But directly it comes to the question of reefing, many of its excellences disappear. Owing to its great proportionate beam, often approaching one-half the length in ballasted boats, and the strong downward pressure of the sail acting on the long lever, these boats all have a keen inclination to bore by the head, and become almost unmanageable when pressed. The single big sail is not a cruising rig, and the absence of forward and after overhangs is a grave disadvantage in boats so liable to great changes of form when heeled. Modern catboats have *some* overhang, but, taking it all round, the type is, for our climate, too short, too shoal, and too difficult to handle.

It is here that the advantage of compromise comes in. Given somewhat less beam, more body, and fair overhangs, a type may be evolved that will be a good seaboat of excellent stability, of a range far exceeding that of a shallower, wider boat, and offering as much room below as any boat of the same length of waterline. A draught of from one-seventh to one-ninth of the waterline length and beam of from .35 to .4 of the waterline length, give a basis upon which to plan a cruising centreboard yacht of a highly satisfactory kind, which will be sturdy enough to face a stretch of open sea successfully, strong enough to take the ground, and of such a draught of water as will enable almost any district, however full of shoals, to be explored. Such boats up to about ten tons displacement would be excellent for almost every description of cruising, both in winter and summer, having sufficient area of lateral plane without the centreboard to enable them to be worked in water only a few inches deeper than their draught. They would ride, without the board, steadily to their anchors in strong winds, when a much shoaler type would sheer about badly unless the board were down, and they might be so designed as to be as pleasing to the eye as the deep keel type, a feature not without its advantages at all times, but especially in the market.

The example given in Fig. 28 is such a craft, offering good accommodation, fair speed, and good seagoing qualities. The chief points to be aimed at in the design of boats of this class are balance of form at moderate inclinations, good clearance aft, long flat buttocks and bow lines, and an easy, flattish rise of floor. The displacement should not be too great, and the area of midsection need not be larger than is shown for the length of waterline on the diagram in Part IV. The sail area required to drive this boat would be about 100 sq. ft. per ton displacement, but this might safely be exceeded by at least ten per cent., with a sufficient crew to handle it. The board should be of wood, weighted so that it would sink easily, the case serving as the centre support for a table. Boats of a similar type are in common use in America, some with more and some with less beam and draught, but such boats are rarely seen in our waters, though for many districts they would be suitable. It has as much stability and safety as the bawley type, which was popular on the Thames some years ago, but in handiness and weatherliness would be found superior. One can only conclude that prejudice has been at the bottom of its neglect, because the very moderate draught of hull would allow it to be used in any district. It cannot compete with the barge yacht in the matter of sitting upright on the ground, but in soft mud it would be likely to sit quite near enough to the upright to be habitable, while on hard ground legs could be used. Unless all experience is wrong, it would be a faster and better sea boat. In the matter of first cost it would be more expensive, but it would always command a better price in the market.

For greater lengths of waterline than 25 ft. the advantages of the type begin to disappear, the board grows large, the draught begins to nearly approach that of a keel boat, and the sail area increases to a point beyond easy handling with a small crew.

But for the length given and decreasing therefrom to about 18 ft. L.W.L. an extremely useful and economic class of boat can be designed, suitable for almost every kind of cruising, and offering quite good accommodation, speed and seaworthiness.

Fig. 29 shows one of the smaller class, a good number of which have been constructed during recent years. This type of boat has been created as the outcome of a desire to enjoy foreign cruising in small

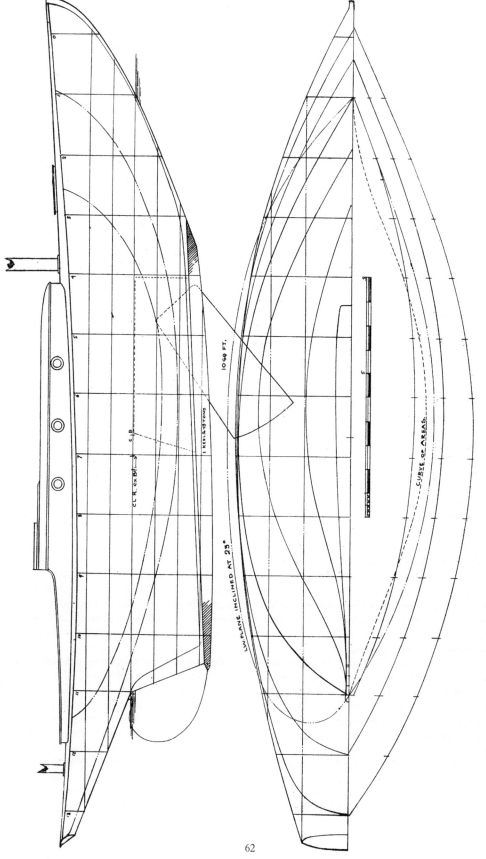

Fig. 28.   L.O.A., 35·5 ft.; L.W.L., 25 ft.; Beam, 10 ft.; Draught, hull, 3·5 ft.; total, 7 ft.; Displacement, 7 tons.

craft, and is the result of gradual experiment on the lines of producing an easily handled and inexpensive boat which could be handily shipped on a steamer, and would give accommodation for two men, while yet being a true single-hander. The short overall length is required by the demands of shipment, as is also the comparatively light displacement. The light draught and the centreplate are called for by the nature of the foreign cruising contemplated, as well as by local conditions. The sharp stern is traditional, and marks most of the boats yet built, though in some instances the " sawn-off " counter and the canoe stern have been adopted. But the shape of the stern has little to do with the fundamentals

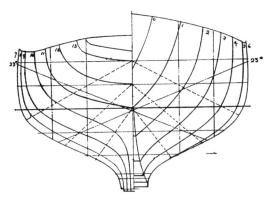

of the type. These reside in the large beam, compact section, centreplate and rig. If in these particulars great variation takes place the type is so far departed from that its unique advantages are in danger of being lost. " Compromise " is writ large on every detail of these little craft, but it has proved effective. The numberless cruises along the coast of England, Holland and Germany that have been accomplished in these boats is a testimony to their ability to face the sea and to navigate tidal waters with success and comfort.

A smaller and lighter type, from 15 ft. to 18 ft. length overall, is still in existence in the Humber Yawl Club. They have been very useful little craft for cruising on inland waters, and many extensive cruises have been accomplished in them abroad and in England. But for tidal waters and large estuaries the boats have been found too small for comfort, and lacking in power to

face the broken, short seas of a weather-going tide, though they make excellent day boats and week-end cruisers, especially the 18 ft. class. The fifteen-footers are only wide canoes of very light displacement, and depend mainly on form for stability. This being so, it will be concluded that their limit of stability is soon reached, and that they require a careful hand at tiller and sheet when sailing.

The question as to the proper placing of the centreboard in relation to the other centres of the yacht is a much-disputed point. Even in America, where the board is so extensively used, no definite rule seems to be in existence. In a very able paper read before the Society of Naval Architects, Mr. W. P. Stephens, the American designer, stated that he " knew of no rule or formula that covers the case," and in the discussion that followed, one of the naval architects present stated that in two vessels of the same hull form and with boards of the same area, there was a difference of four feet in the fore and aft position of the boards, and that the vessel with the board placed furthest forward was in all respects a better performer to windward. In my own experience most of the failures I have met with and been consulted upon, were due to the position of the board being too far aft, and I am inclined to believe that in craft having some amount of rake to the keel the board can scarcely go too far forward unless the vessel is of very extreme beam. When it is remembered that the speed and the consequent pressure on the lee bow is less when a vessel is close hauled than it is when she is reaching, it will be realised that there is far more likelihood of failure if the board is too far aft than if there is a slight error in the other direction, and it is an undisputed fact that unless the C.L.R. is brought appreciably forward by the lowering of the board, the advantage of the added few feet of additional area of lateral plane is almost inappreciable.

When very moderate draught is required, but the use of the centreboard is inadmissible, the type of midsection used may contribute materially to success or failure. It may be taken as an axiom that some types of midsection are in themselves leewardly unless of great depth. It is noticeable that the Falmouth quay punt has a very different set of sections from the Itchen Ferry boats,

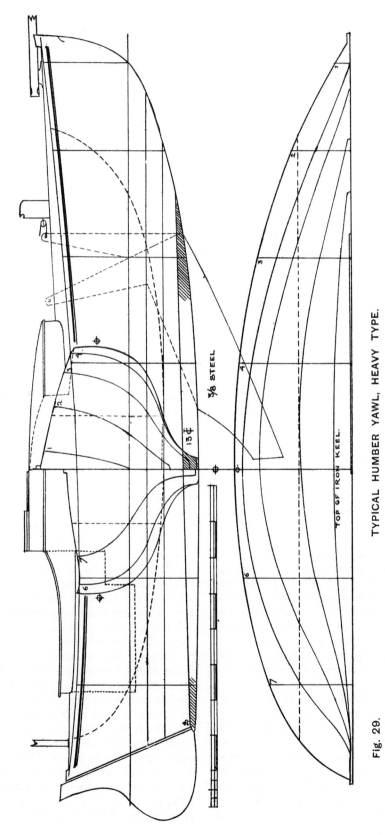

Fig. 29. TYPICAL HUMBER YAWL, HEAVY TYPE.

L.O.A., 21 ft.; L.W.L., 19.5 ft.; Beam, 6.9 ft.; Draught, 2 ft.; Displacement, 2 tons; Sail Area, 320 sq. ft.

and that the latter have, as a rule, a greater proportionate beam for their length, but that their average draught of water is less. Some early quay punts had very sweet ogee midsections not at all unlike those of the Southampton boats, but they rarely had so great a beam, that is, in the working boat class, but they had greater draught. The earlier bawley boats on the Thames had also ogee sections with great beam, so it will be concluded that light draught demands great beam and a midsection with a considerable amount of curve in it. That is, the boats of light draught demand a higher centre of buoyancy than

the open sea, or in the tideless lochs of Scotland, the light draught boat, unaided by a centreboard, and even if furnished with one, cannot hope to compete with the long-heeled type, and the harder it blows the less able is it to make a showing. I have, in Scotland, seen boats of 17 ft. W.L. drawing 6 ft. aft, and 19 ft. boats with over 7 ft. draught. Such boats would completely outsail to windward boats of lighter draught in their own waters, but the difference in weatherliness would be greatly less in tidal waters. The fin-keel boat, owing to the extensive flat surface of the metal or wood fin, holds a good wind with a **very**

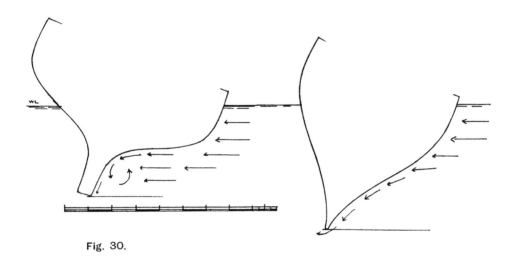

Fig. 30.

the deeper craft, which, being further interpreted, means a boat *initially stiffer,* a boat that must be sailed as upright as possible if the best is to be got out of her. Because every degree of heel so lessens the area of lateral plane that it soon becomes ineffective as a preventer of leewardness.

A flat side to the keel and a high curve in the garboards does hold a boat to windward extremely well. The reason is shown in Fig. 30, which illustrates approximately the action of the water about the region of the keel in the two different types of section.

In all waters where there is considerable use to be made of tidal currents, draught of water may be considerably reduced. A weather-going tide of three knots and over will bring an extremely shallow boat to windward notwithstanding leeway. But in

moderate amount of draught, and is weatherly. Yet this type is naturally deficient in body and head room, unless a high cabin top is carried, so that what is gained in one direction is paid for by a loss in another.

One other point must be noted with regard to beamy boats of light draught. As the angle of heel becomes more pronounced the greater the " ardency," or the tendency to press towards the wind, becomes. If the boat is well designed this should be so, as any light draught boat that does not exhibit a tendency to " pull " will not be a good performer to windward, and the more she is heeled the greater should be the pull. Those of us who are old enough to have sailed in the long narrow " 1730 " type will remember that this was never the case with

¹ them. On the contrary they often required lee helm, and a lot of it, when pressed hard in running or reaching.

Many modern racing boats with round forward lines and cut away keels take lee helm at small inclinations and low speeds, but at large angles of heel and high speeds require a great amount of weather helm. To those who remember that a round bow causes the centre of resistance to travel forward as the angle of heel, the speed, and the lee bow wave increase, will not be surprised at this statement, but will perceive the reason for it. It illustrates the difficulty in approximating the distance that should be kept between the C.L.R. and the C.E., and shows how large a number of factors have to be considered in deciding on this important point. But it is always better to have the balance on the side of ardency, as correction can be more easily made by disposal of ballast and sail centres than when lee helm—always dangerous as well as detrimental to windward work at speed—is the trouble to be remedied in shoal-draught craft.

1 The reference is to the very narrow, deep, 'plank on edge' yachts designed to take advantage of the YRA 'Tonnage' rule:

$$\frac{(\text{LWL} + \text{beam})^2 \times \text{beam}}{1730}$$

It was superceded in 1888 by the 'Length and Sail Area' rule.

The Design and Construction of Small Cruising Yachts.

BY

ALBERT STRANGE.

VI.—CONSTRUCTION.

THE actual construction of a yacht is, as a rule, entirely beyond the power of an untrained amateur, unless the yacht is a very small one, and the amateur is uncommonly good in the use of tools, especially of that fearsome weapon, the adze. It is therefore not proposed to offer instruction in the art of building, but to explain processes and methods of work, sizes of scantlings, and the nature of materials, so that the budding designer may have enough knowledge to keep him fairly right in making his construction drawing. No one who has not had an intimate acquaintance with the actual work of building—who has not at some time or other hardened his hands by manual labour, and who has not watched the building and the breaking up of yachts and other vessels, can hope to gain much knowledge from any written description of the processes of construction.

When the builder receives the plans of a yacht, he should also be provided with a table of offsets. These are the measurements of all heights, widths and depths on the drawing—and are used in conjunction with the plan in "laying down" the design on the mould loft floor. Laying down means drawing the yacht full size, so that patterns of the frames, or moulds, keel, stem and sternpost can be made. The table of offsets is a check against any errors of measurement both by the builder and designer. Errors will creep in in scale drawings, but they creep in in greater numbers if there is no table of offsets to refer to, because in an inch scale drawing the bare thickness of a line means quite an appreciable size when magnified by twelve.

Formerly all drawings were made to the outside of frames without the planking. Nowadays the drawings are to outside of plank unless otherwise specified, and the builder deducts the thickness of plank on the floor. It is the better plan to allow him to do this, unless he is unused to the work, which sometimes happens when he has only built " by eye " such craft as small fishing boats and rowing boats. With such a builder the designer is bound to give a good deal of assistance in laying off if he desires his work to receive full justice. It will also be probable that he will have to supply and explain a great number of sketches of detail. All this takes up a lot of time, and therefore the cheap rule of thumb builder is not so cheap as he seems to be at first glance. A good many of these men, however, take a real pride in their work, and it is a pleasure to assist them. Some, on the other hand, are difficult to persuade that there are various ways of doing things, some of which are a good deal better than others.

To anyone unversed in the details of construction there are good guides as to scantlings in the books issued by Lloyd's Register, which will serve in most cases all the requirements of the amateur. The professional yacht builder or designer has, however, his own methods, sizes and arrangements, and to those of good reputation much may be trusted. Of course, if a yacht is to be built to Lloyd's Classes, the scantlings and fastenings must conform to the rules laid down, and the yacht must be inspected periodically at each stage of construction by one of the Society's Surveyors.

It is often the case that amateurs believe that no construction can be too solid or heavy to meet all the emergencies of the

the ground in dry harbours, extra strength is needed, but it may be taken as certain that no cruising yacht need be built as strongly, or, I should say, as heavily, as a fishing boat, because they have not usually to "rough it" to the same extent. Heaviness does not necessarily mean strength, and the strains experienced by cruising

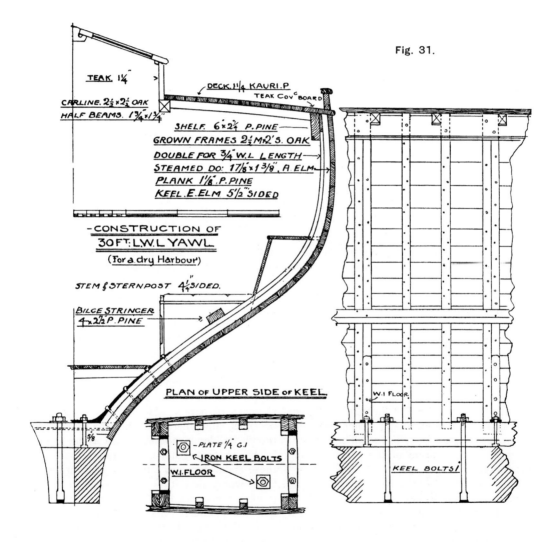

Fig. 31.

TEAK. 1¼"

DECK. 1¼ KAURI.P
TEAK COVᴳ BOARD

CARLINE. 2½"×2½" OAK
HALF BEAMS. 1¾"×1¼"

SHELF. 6×2¼ P. PINE
GROWN FRAMES 2½"M×2"S. OAK
DOUBLE FOR ¾ W.L LENGTH →
STEAMED DO: 1⅞"×1⅜". A. ELM.
PLANK 1⅛.P.PINE
KEEL.E.ELM 5½"SIDED

-CONSTRUCTION OF
30 FT. L.W.L YAWL
(For a dry Harbour)

STEM & STERNPOST 4¼" SIDED.

BILGE STRINGER
4×2½ P. PINE

PLAN OF UPPER SIDE OF KEEL.

W.I FLOOR

⅝

-PLATE ¼ G.I
IRON KEEL BOLTS
W.I.FLOOR

KEEL BOLTS

sea, and many owners, too, seem to take a pride in scantlings that are needlessly heavy, and occasionally insist on their ideas of solidity being carried out—regardless of the fact that every pound of unnecessary wood means one pound less of the necessary lead or iron ballast. In special cases, such as where the yacht has constantly to take

yachts under normal conditions are much less than those of a racing yacht. The hard "carrying on" of racing is what wears racers out and pulls them out of shape. Ordinary cruising never tries a yacht to the same extent. The chief source of weakness in cruisers is not light scantling, but inadequate connection of all the parts that receive local strains, and lack of close and accurate fitting of all parts, irrespective of caulking. Masses of deadwood fore and aft, so common thirty years ago, are no addition to strength, but one of the chief causes of rot and decay. Moreover, they punish the boat in a seaway by making her pitch and scend owing to weight in the

that a yacht of 28 to 30 ft. waterline should have at least a couple of stringers below the shelf if steamed frames only are used, especially if there is a good deal of hollow in the section. Otherwise the boat is apt to lose shape under the stress of sailing. Motor cruisers up to 40 ft. waterline may be satisfactorily built on steamed frames,

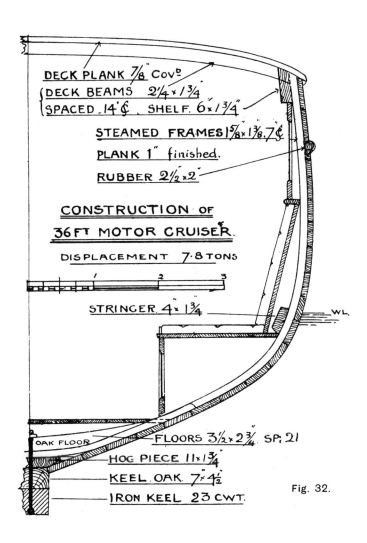

DECK PLANK 7/8" COVᴰ
DECK BEAMS 2¼ × 1¾
SPACED .14 ¢ , SHELF. 6 × 1¾"
STEAMED FRAMES 1⅝ × 1⅜ .7 ¢
PLANK 1" finished.
RUBBER 2½ × 2"

CONSTRUCTION OF
36 FT MOTOR CRUISER.
DISPLACEMENT 7·8 TONS

STRINGER 4 × 1¾

W.L.

OAK FLOOR
FLOORS 3½ × 2¾ SP. 21
HOG PIECE 11 × 1¾"
KEEL OAK 7 × 4½"
IRON KEEL 23 CWT.

Fig. 32.

ends. Since the introduction of the International Rule, with its fixed scantlings, a good deal of modification has been made in yacht construction. Yachts up to 33 ft. waterline have been built with all steamed timbers, and a good many different combinations of "grown" and steamed framing have been allowed and used with varying degrees of success. It may be remarked

if they are well provided with wood floors and stringers of good scantling.

Fig. 31 illustrates the construction of the 30 ft. yacht given in the article on design (Fig. 25), and Fig. 32 shows the ordinary scantling for a motor cruiser of 36 ft. length. The term "siding" means the thickness from flat to flat on a frame (or other member), "moulding" means the thickness from curve to curve. It is difficult to work a steamed timber of greater moulding than 2 in., or having a sectional area of over 4 sq. ins. In fact, it is better if timbers of this sectional area are required to have the siding a little more and the moulding somewhat less, both for ease of bending and strength of fastening, as the plank nails may be placed more out of a straight line. A steamed timber of

Frames are often doubled only to the turn of the bilge or thereabouts, and when this plan is adopted the joints can be easily arranged to come abreast of a long length. Top weight is saved, but the frames should be double throughout where the chain and runner plates will come. Steamed frames should always be at least 25 per cent greater moulding than the thickness of plank they support.

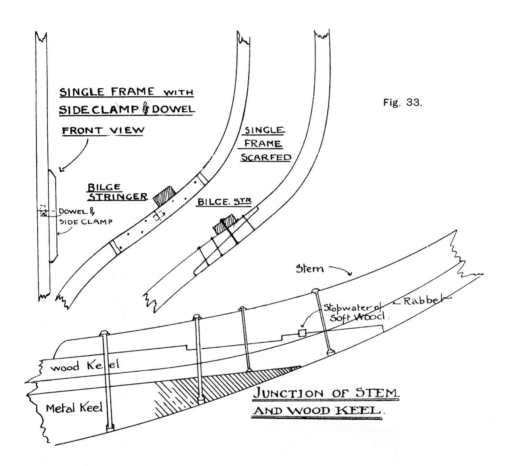

Fig. 33.

2.4 in. siding by 1.7 in. moulding gives the same strength, bends easier, and fastens better. On the other hand, a "grown" frame, especially if single, is better if the moulding is rather in excess of the siding. Unless in small sizes of yacht, grown single frames are not to be recommended. They have to have a joint somewhere, except when fitted in the bows, and the side clamp where the joint comes—or the scarph —if no side clamp is used, is a source of weakness. (Fig. 33.)

Iron floors, worked across the keel and connecting the heels of the timbers, are now almost exclusively used in craft of any size. The ordinary wrought iron floor is not very strong unless thick in the "throat," or angle. But they are, on the whole, cheaper and less difficult to fit and fasten than the steel angle plate sometimes used. Wooden floors occupy so much space that they are seldom used in small yachts; at any rate, in the vicinity of the cabin. In larger yachts they are still used, but should always be worked with a stout keelson on top, through which a strong bolt is driven and clenched, from the underside of main keel to the top of the keelson. In America a simple piece of thick oak plank is often used as a floor, fitted and fastened to the sides of the frames. It is not a very satisfactory method, but as crooks are not easily obtained there, and wrought iron floors not always available, it is a fair substitute.

It is particularly necessary to galvanize all ironwork used in yacht construction. This should not be done until the work is accurately finished and fitted, and all holes bored. The chemical action of lead and bilge water combined is very destructive to ungalvanized iron. The action of clean salt water and lead is also dangerously rapid,

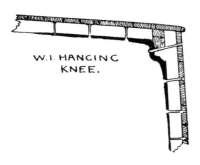

W.I. HANGING KNEE.

and iron bolts should never be used for lead keels. These should be fastened on by means of gun metal bolts with large heads and screw nuts and washers. The lead or iron keel bolts are not usually driven through the iron floors, and are better if well clear of these. [1]

The use of iron fastenings, even when galvanized, is to be deprecated in any yacht work. The difference in cost is very slight, and in the course of time iron fastenings are certain to show rust, as in driving them all the zinc is knocked off the heads, and they at once begin to rust, even when well stopped over. It is true that iron is stronger, size for size, than gun metal, and far more easily driven, but rust is very unpleasant to look at, and impossible to conceal. It also ruins all wood in its neighbourhood. Some small builders have a very objectionable practice of fastening the plank ends in stem and sternpost with iron nails, "to prevent the planks springing off," they say, as newly-driven copper nails do

not hold well unless clenched. But good "composition" nails hold, and if copper nails are roughed on their angles and driven diagonally they too will hold. I have seen many boats quite ruined by this pernicious practice, which should be carefully prohibited in the specification and watched for in the actual building. Screws, when carefully driven, hold plank ends excellently. Copper fastenings should be either "turned," that is, the end of the nail bent back and hammered into the wood, or "clenched" over small round washers, the nail head being supported by a hammer while both operations are being performed. In small vessels all fastenings through both grown and steamed timbers should be clenched. In larger work of from 40 to 50 ft. W.L. at least fifty per cent of the frame fastenings should be clenched, the remainder being "dumps" or fastenings which do not go quite through the frames. Dumps are small round bolts with solid heads, and are of the same thickness throughout. "Spikes" are shaped like ordinary nails, of yellow metal, or composition, and are not so good as dumps, which are best of copper.

In the South of England the general practice in small work is to clench all through fastenings. In the North and in Scotland, the nails are often turned only. If properly done this is quite a good holding fastening if care is taken to avoid turning the nail points in exactly the same line as the grain of the timbers, otherwise the timbers are often split.

All wood and iron floor fastenings should invariably be clenched, as also fastenings of scarphs in stem, keel, and the framework of the counter.

As to sizes of materials, plank timbers, frames and posts, these vary slightly in different localities, but a good rough working rule is as follows: Stem and stern post four times the thickness of plank, for siding, five times the thickness for least moulding.

Steamed frames, one and three-quarter times the thickness of plank for siding, and one and a quarter times the moulding, and spaced about five times the siding on centres. Planks should, in thickness, be $1/32$ in. for each foot of waterline length, plus $4/32$ in. Thus for a 24 ft. W.L. boat: $24/32 + 4/32 = 28/32 = \frac{7}{8}$ in., and a 40-ft. W.L. boat: $40/32 + 4/32 = 44/32 = 11/8 = 1\frac{3}{8}$ in., but the plank

1 See page 76 for footnote.

for boats of about 18 ft. W.L. should never be less than ⅝ in. finished. The figures given allow a good margin for planing off.

As a rough working rule this will be found a good guide when the plank is of good quality pitch pine, oak, or hard mahogany. Teak may be slightly less. In all woods the planks should be narrow, not exceeding 4 in. or 5 in. in the widest part, except the garboard strakes. The accompanying table gives the spacing of frames and sizes of fastenings sufficiently strong for general cruising requirements. A bilge stringer should always be worked in all vessels down to about 25 ft. W.L.

The thickness of deck planking should not be less than that given in the table if the deck is single thickness. The plank should not be more than 2½ in. wide (2 in. is better in the sizes from 25 ft. W.L. downwards). Deck planking of the smaller yachts should be kept varnished, as the exposure to wet and dry, and sun and spray, makes it very difficult to keep tight decks when there is only one inch thickness. Narrow planks and good caulking are essential. If lighter decks are desired they should be covered with canvas and painted, and the deck beams spaced a little closer to avoid spring.

The modern practice of " swept " decks —where the planks follow the curve of the sides and butt into a centre king plank— is not to be recommended in beamy boats, as tapered decks keep their place better. The great pressure necessary to get the planks round the deck curve of a beamy boat often cracks the planks across, and the caulking forces them off the beams. There is no doubt that straight-laid decks are strongest and most practical, but fashion is against the look of them. " Secret " fastening is not so effective in holding as the straight-driven nails with dowelled heads, but here again fashion forbids. Deck planks should be laid with the grain on edge.

The greatest weakness to be guarded against is the forcing out of the yacht's sides by the deck caulking, and tie-rods should always be fitted where deck openings occur. These are driven through the head of frame and through stringer and carline, and clenched. They should be at least ⅜ in. diameter yellow metal rods. Where the planks are fastened to the beams across the

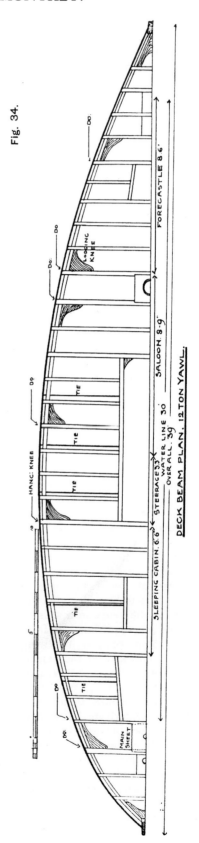

Fig. 34.

DECK BEAM PLAN. 12 TON YAWL.

whole width of the deck they are not wanted, of course, but in all breaks of deck, where cabin-tops, skylights and large companions come they are very essential to strength and watertightness. (Fig. 34.)

Hanging knees connect the beams to the shelf and timbers, and are always of galvanized iron. Lodging knees connect the

become leaky. Especially is this the case about the covering boards and in the region of the rigging and runners. The result is not only discomfort to the crew, but sure decay and rot in the frames and planking of the vessel. Proper design and construction in this respect, though it adds, of course, to top weight, is far better than

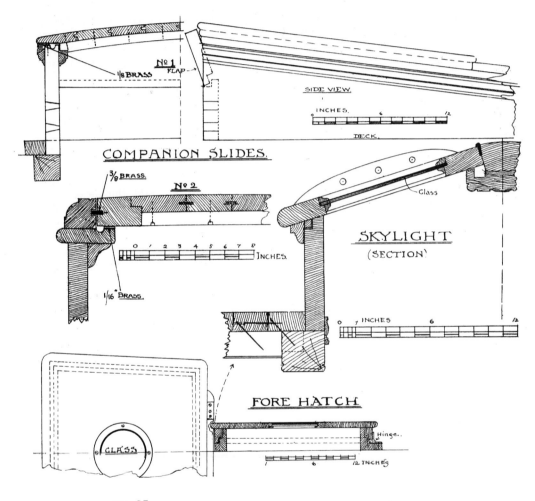

Fig. 35.

beams with the shelf, and provide against diagonal stresses. They may be of iron, but are usually, and preferably, of oak crooks, well fastened by clenched bolts. It is the regrettable practice of many small builders to place too few of these essential connections between deck and hull in their vessels. The result is that the yacht soon pulls out of shape, and her upper works

the use of extra heavy scantlings of frame, plank and deadwood, which, of course, add much more to weight without giving greater strength.

The sketches in Fig. 35 show some usual methods of the construction of companion and skylights, and Fig. 36 shows two methods of construction of the canoe stern and rudder attachments. The steel tube and rudder stem are to be recommended in yachts of the smaller classes, as being lighter and less complicated than wood construction. Where wood is used as a rudder casing it should be of teak. As this part of the yacht is apt to be neglected, and also to suffer from lack of ventilation, rot very often attacks the rudder trunk when made of other wood, even of oak.

Fore companions in the larger sizes should be made on the same principle as main companions, but of rather less dimensions; 22 in. clear is a good width, and is often as large as the deck space allows. In small craft there is rarely room for the companion to slide, so it has to be of the hinged sort. It should never be made unattached to its coaming, as it is not an uncommon occurrence for an unattached cover to be lost overboard through carelessness. Circular covers are pretty to look at, but not the best for actual practice. The one shown in Fig. 35 gives a form less liable to leakage than many, but trouble in this respect is often experienced in bad weather with any shape of hatch.

The question of a windlass of some sort becomes urgent when the yacht reaches about eight to ten tons displacement. It is often the best economy to have one specially made. Some of the small windlasses on the market are of very poor design and power, and the necessity of having to stoop or kneel takes away a good deal of power from the worker. A well-designed mast winch will often prove extremely useful in breaking out the anchor from a stubborn bit of blue clay, as one can, from its height, put a good deal of power into it. But it should have a *very* strong spindle.

If a windlass is decided on it should be so made as to receive the heel of the bowsprit, and thus save the weight of bitts. Of course, no racing yacht of eight or ten, or even twenty tons, would have a windlass; the chain is always got by hand, assisted by a powerful purchase to hook into the links when the strain of breaking out has to be met. A simple steel shoe takes the heel of the bowsprit, this shoe being fitted with various cleats strong enough to take the chain and bobstay purchase, etc.

Provide, if possible, a chain locker as near the mast as it can be placed, and see that it is easily accessible. Chain, unless

carefully tended, has a way of curling itself into twisted knots, which prevent it from running through the hole in the cover of the locker.

The junctions of stem and deadwoods with the keel about the region of the rabbet are kept watertight by means of a small plug of soft wood driven transversely when the timber is bolted in place and before the plank is put on. The soft wood swells, and thus prevents water from passing. The heel of the stern post is tenoned to the keel, the old plan of a half scarph not allowing a bolt to be driven upwards into post and deadwood, as it only forces the scarph apart, and is thus lacking in strength. (Fig. 36.)

In planking, care should be taken that the seams are not too narrow. The plank should just touch on the inside, and gape slightly on the outside. This allows the caulking to be retained. If the plank is very dry and of slight thickness it is apt, if too closely fitted, when the yacht is put afloat, to swell so much as to buckle off the frames, or to split at the nails. A refinement of caulking is adopted in at least one first-rate yard by simply inserting narrow, wedge-shaped strips of clean wood between the planks. These wedges are fastened in by means of a special waterproof glue, the composition of which is a secret, but it is probably a solution of india-rubber and shellac in naphtha. This plan makes a beautiful finish from the first, as the caulking is not squeezed out when the plank gets wet, as is usually the case, for a season at least.

When a yacht is constructed under cover (as it should be) it is an excellent plan to give all underwater parts several dressings of paraffin oil. This seems to set the planking in its place, hardens it, and gives it great power of resisting decay. Whenever I have been able to induce builders to adopt it, after examination has shown its benefits, and I have no doubt that many cases of dry rot could have been prevented by its use. The oil is completely absorbed by the wood, and paint or varnish goes on well afterwards. It is specially valuable for boats built to light scantling, and double-skinned boats. One of the latter, which received five dressings of oil twenty-three years ago, still shows no sign of decay, though during her existence she has been hard worked, and at times none too well cared for.

In craft up to 35 ft. W.L. the deck beams may be half dovetailed into the shelf and the top of beams worked flush with it. In this case the depth of the shelf should be not less than $2\frac{1}{2}$ times the thickness of beam at centre.

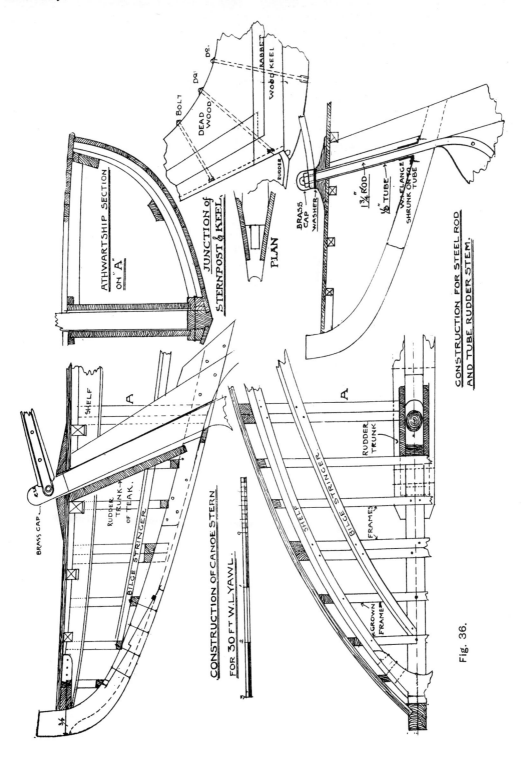

Fig. 36.

SHEILA
BECALMED
IN
LOCH FYNE.

ROBERT.E.GROVES. 1908.

YM June 1908

---

1 From page 71

Although the electrolytic corrosion problems caused by the use of dissimilar metals had been recognised since the 17th century (when iron-fastened ships were first sheathed with copper), electrolysis was not fully understood. Contemporary practice was to bolt or rivet iron floors with copper or yellow metal, and this was approved by Lloyd's. Even if galvanised-steel fastenings are used, the floors themselves often touch the copper fastenings which hold the plank to the frame beneath. The resulting corrosion not only speeds rusting of the floor, but also breaks down the structure of the wood. A simple expedient to arrest, or at least slow down, this process, which has been the downfall of so many wooden hulls, is to place a neoprene, or other synthetic rubber, insulating strip between the floor and the frame to which it is fastened. In Strange's day, tarred felt, though not as effective, would have helped.

The Design and Construction of Small Cruising Yachts.

BY

ALBERT STRANGE.

VII.—BALLAST, SAIL PLANS AND SPARS AND RIGGING.

BALLAST.

THE amount of ballast in proportion to total displacement varies from .35 to .5 in cruising yachts of modern type. This variation depends upon the shape and construction of the yacht. In the deep, narrow type popular twenty years ago, when stability depended almost entirely upon weight at great depth, an even greater proportion than .5 was occasionally used, and in racing craft very commonly used. Even to-day small racing craft of less than four tons displacement carry lead keels of over two tons, but it must be borne in mind that cruising yachts have necessarily to carry what practically amounts to a small cargo of cruising requisites, *viz.*, stores, furniture, stoves and utensils, water, warps, spare anchors, etc., which all go towards reducing the proportional weight of lead or iron on the keel or below the cabin sole.

A wide, shallow, strongly-built hull of the centreboard or fin-keel type cannot carry, and does not require, so great a proportion of ballast to displacement as a deeper-drawn vessel, as the constructional weight of hull, cabin top, spars, etc., is greater in proportion than it is in a deeper-bodied, less beamy vessel. A yacht of 25 ft. L.W.L. by 10 ft. beam and 4 ft. draught may be of the same displacement as one of 25×8×5, yet, her hull being so much larger, would be heavier, and thus her proportion of actual ballast would be less than that of the narrower boat. But, being so much wider, her centre of buoyancy shifts much quicker and further to leeward when heeled, and thus her righting couple is, especially at small angles of heel, appreciably longer. As the angle of heel increases this length diminishes more or less rapidly, until at a given angle, say, about 35 or 40 degrees, it may disappear, whilst in the narrower but deeper yacht the righting lever is still long enough to ensure complete stability at a much greater angle of heel. This initial stiffness of the beamy boat is an asset of great value in cruising under ordinary circumstances, but it also tends to quick and jerky rolling, and sudden and violent recoveries from an inclined position back to the vertical, which are uncomfortable to the crew and dangerous to spars and rigging when cruising in deep offshore waters. For ordinary inshore and estuary cruising these disadvantages are so rarely felt as to be almost negligible.

The amount of ballast that may be placed below the keel should not be more than about .35 of the total displacement, if easy behaviour is required in a vessel of deep draught. The *whole* of the ballast should never be outside, especially if the ballast keel is of iron, for the reason that, however exactly calculated the weight of keel may be, any error that may appear in trim when the boat is afloat is difficult to rectify, and no one, however learned he may be in naval architecture, can calculate or foresee in what sort of trim the vessel will sail and steer best. She may require to be more by the stern, or by the head, than as originally designed, and can only be trimmed to her best sailing line by the addition of inside ballast over and above the calculated amount, and her displacement and draught are thereby increased. Sometimes a yacht does best with less weight than was anticipated (though this is a rare occurrence), and if the whole of the ballast is outside and of iron, the removal of part of it is a costly and difficult job. Provision can be made, even in iron keels, by leaving spaces at the ends for insertion or removal of blocks of wood or metal, but the better plan is to have the iron keel of such a weight as to allow of at least ten per cent of internal ballast to be used. This can be placed as required to bring the vessel to her designed trim, and subsequent alterations to trim are then easily made by trial when under canvas.

Iron keels of from two to six tons are generally cast in one piece. Beyond this weight it is often difficult to obtain them without a scarph. Should this be anywhere near the centre of the keel its strength is much reduced, and one of the great advantages of an iron keel is the fine foundation of strength it gives to the vessel. If a scarph must be used it should not be less than one-fifth of the total length of keel. Short scarphs are utterly useless from the point of view of strength. The bolt holes should be drilled in the scarph to obtain exact fitting, and in the other parts of the keel where they are cored they should also be cleaned out with a drill. A close driving fit for the bolts is of importance to prevent rust, and the heads of the bolts should be well above the lower level of keel and filled with good cement.

Cast iron varies much in weight, according to the quality of metal used. The common weight of cast iron is 450 lbs. per cubic foot, but iron weighing no more than 430 lbs. is not infrequently met with in country foundries, owing to the fact that old furnace bars and any old rubbish of iron are often added in large quantities to a small amount of good pig. The lower part of the wood keel and the upper face of the iron keel should have a thick coating of stiff tar before the iron keel is bolted on. Though not often is it the case that the iron keel is fastened in place before the rest of the building is proceeded with, it is much the better plan, and should be adopted if

possible. When this is done, the iron floors must be fastened to the wood keel by means of large stud bolts, or coach screws with hexagonal heads. When the iron keel is fastened on after the hull is finished, the floor bolts are ordinary square-headed bolts with screw nuts, to hold the floors down. These are driven from the underside of the wood keel before the iron keel goes in place. The iron keel bolts should have big washers under the nuts on the upper side of wood keel. [1]

The following table gives the usual load per bolt for iron keel bolts :

Diameter of bolt, $\frac{5}{8}$ in., load per bolt, 270 lbs. ; $\frac{3}{4}$ in., 420 lbs. ; $\frac{7}{8}$ in., 550 lbs. ; 1 in., 750 lbs. ; $1\frac{1}{8}$ in., 850 lbs. ; $1\frac{1}{4}$ in., 1150 lbs. (not more than 1 ft. 6 in. apart).

The load for Muntz or yellow metal bolts should be fifteen per cent less. These should always be used for fastening lead keels. Lead weighs 710 lbs. per cubic foot, and is thus a superior metal for ballast keels so far as its specific gravity goes. There is, however, little strength in a lead keel, and it is easily damaged and occasionally bent by accidental grounding on rocks or irregular hard ground. Apart from this drawback its surface always remains smoother than that of iron, and where the draught of water is strictly limited a keel of smaller dimensions for the same weight as iron can be used.

Internal ballast should always be of lead. It is cleaner, and, apart from its first cost, more economical in the long run than is iron, as it can always be disposed of in pieces of manageable size at about the market price of lead. Lead keels, being of larger size and not easily manipulated, realize less than pigs, as they have to be cut up into pieces of reasonable size before they can be re-cast or placed on the metal market, and the cutting up is a costly business.

Cement ballast, however advisable for old boats, should never be used in new craft. There is no necessity for its employment, and, in case of repairs being needed, its presence is very awkward. In the extreme ends of the yacht (about the lower part of the stem, and in the vicinity of the junction of timbers with the deadwood aft, it is often used, and in these places it does good in keeping out dirt and damp, and making a fair channel for the drainage of water to

1 See page 86 for footnote.

the pump well. But in all cases it should be insisted on that all surfaces of wood intended to receive cement should have three or four good dressings of tar and paraffin before the cement is applied. Otherwise rot may soon set in. The cement should be in the proportion of three of cement to one of fine sand, not three of sand to one of cement, as is often the case. Cement is largely used in the bottoms of fishing boats, but there is good cause for it here. The slime and dirt arising from the storage of fish in the hold could not be properly cleaned out unless a flat cement floor were made. In a yacht dirt is always out of place and always avoidable.

The centre of gravity of an iron or lead keel should be very closely under the C.B. when internal ballast is used. When a motor or heavy tanks are placed abaft the C.B., then the C.G. of the keel is better placed ahead of the line of C.B. The yacht can then be trimmed to her designed W.L. by internal ballast.

When it is desired to place the whole of the ballast outside, the position of the centre of gravity of the hull without the ballast must be ascertained by an exact calculation of the weight and C.G. of every piece of material used in her construction, a very long and tiresome piece of work, which must be carefully and accurately done, or the result will be useless. Having ascertained the exact weight of all the material— including mast, sails, anchors, etc., this weight must be subtracted from the calculated displacement, which will give the weight of the lead or iron keel. The position of C.G. of keel is then found as follows :

Example—A yacht of twelve tons displacement has seven tons weight of hull, etc., leaving five tons for weight of lead keel. The C.B. is 16 ft. aft of stem; the C.G. of materials is 15.5 ft. aft stem :

Total Displacement—

12 tons × distance of C.B. (16 ft.) = 192

Weight of Hull and Equipment—

7 tons × distance of its C.G. (15.5) = 108.5

Divide by weight of lead keel = 5) 83.5

16.7

16.7 ft. is the distance aft of stem for C.G. of keel.

SAILS AND SAIL PLANS.

For any given area the most effective form of sail plan from the point of view of speed alone, is the sloop, and very near to it is the cutter; especially when there is a large proportion of the area in the mainsail. The least effective rig from the speed point of view is the ketch, where the main body of sail is in the centre of the plan, and the remainder of the area adjusted at the fore and after ends of the yacht to obtain a balance. The schooner rig is not very effective as a small yacht rig, though there are good grounds for stating that some advantage in windward work lies with it as against the ketch, as the largest sail is aft. So far as total weight of spars and rigging is concerned, the advantage is on the side of the cutter, and weight of spars and rigging is a factor of great importance in its bearing on stability.

The yawl rig, provided it is so planned that the yacht will handle under mizen and headsails only, lies midway in point of speed between cutter and ketch. It is a favourite rig with cruisers of all kinds, from the humble owner of the canoe yawl to the lordly possessor of an eighty-ton cruiser. The question of its superiority as a useful cruiser rig has long been discussed and disputed. Those who have a longing for speed will have none of it; those who dislike a long boom and a large mainsail to handle, but yet do not wish to go too gently over the waves, say that in a hard breeze the rig is easier on a boat than the cutter. The cruiser who is never in a hurry and who likes a large boat to be handled by a small crew swears by the ketch, and there can be little doubt that from the point of view of comfort and economy he has the best of the argument.

A cutter of 25 tons requires three men. A ketch of 35 tons can be worked by the same crew. A yawl of 35 tons must have at least three men and a good boy. Plenty of 20-ton ketches carry only one paid hand when the owner is a good man at the tiller and a seaman of sorts. But the cutter of 20 tons would be hard work for two men with a good owner if lengthy cruising were indulged in. In small craft, such as single handers, canoe yawls, and such like, the yawl rig is generally the most popular, except in the very smallest sizes, where the sloop rig offers more advantages in speed, and is as easy to handle.

So far as effectiveness goes in what was once considered the crowning proof of sailing ability—windward work—the ratios are somewhat as follows : Given three hulls of equal size and form and 1,000 square feet of canvas, the cutter will get as far to windward in ten hours as the yawl will in eleven, and the ketch in twelve or a little more. But as sailing is seldom always a turn to windward and the intrusion of the motor is of considerable assistance in turning to weather, the time-honoured standard of excellence has to be modified. There is still, however, the very important consideration of wages to be remembered, and it is highly probable that the future will see even a larger number of ketches and yawls aided by small motors than the past has done.

Rig often influences accommodation, especially in the smaller sizes. The position of the mast can be further forward for a yawl or ketch than for the cutter, especially in a long-keeled boat, and this often assists cabin-planning, most of all where the necessary headroom demands a cabin top. It is true that even yet one occasionally sees a cutter's mast projecting through a raised cabin top, but it is equally true that no cabin top thus treated will ever refrain from working and leaking, even when the mast beams are carried across under the raised top.

The apportionment of sail to hull is a very difficult question, and it is hard to find a formula that will fit even a majority of cases, so much does form and type influence the amount to be placed over any length of waterline. It is very largely a matter of experience to exactly fit power, speed and handiness into a perfect combination. A general rule used in France and America is to determine the amount to be given by a fixed ratio of square feet of sail to area of midsection, but even this is only safe when used in comparison with other boats of similar form and ballast plan. To use this formula in connexion with a cruiser of about 30 ft. W.L. and a midsection of 22 sq. ft. the ratio of 40 would give an area of 880 sq. ft., quite a sufficient cruising area for English waters. For Mediterranean and American cruising a ratio of 45 to 50 would be more suitable, but too much for home waters, unless the yacht has a low centre of gravity and plenty of beam. The ratio

of 40 gives practically the same thing as the square of the waterline length. In the case of the 30 ft. W.L. boat this would be 900 sq. ft., and for a boat of 20 ft. W.L. 400 sq. ft. If this is taken as the extreme amount allowable, including topsails and big jibs, it is not a bad guide for a cruiser's trousseau, and may be modified in accordance with the type of boat, amount and position of ballast, and proportion of beam to length. It corresponds, roughly, to 100 sq. ft. per ton of displacement, but only in the smaller sizes of cruisers. Of racers I will not speak, for 650 sq. ft. to about four tons displacement and 20 ft. waterline seems the approximate present allowance, coupled with a beam of about 5.5 ft. The sail is not *carried* in such boats except in light airs.

Effectiveness of a given area is only secured by perfect cut, setting and trim of sails. The present method of cutting and making sails, where the cloths are at right angles to the line of leach, is a very great improvement in all respects on the older method, where the cloths run parallel to the leach. There is a great difference between the propelling powers of the two methods of cut, and, if properly treated, the new style of sail will retain its effectiveness for a longer time than the old, especially if the canvas is of a reasonable substance. Many sails are made of cloth far too light in weight, which soon pulls out of shape. The gain supposed to be derived from lightness is far more than offset by failure of driving power, due to loss of proper curvature from luff to leach. All, or nearly all, modern yachts, even of cruiser type, have the mainsail laced to the boom. Where a patent reefing gear is used it is, of course, essential that the sail be laced. A laced sail also needs a less heavy boom, as the strain is better distributed. But the question as to which has the better driving power, loose foot or laced foot, is one that is not yet finally settled. A good many experiments with models have seemed to show that the loose-fitted sail is a better driver, and it seems certain that in beamy and full displacement boats a loose-footed sail suits better than a laced sail. It is highly probable that no sail yet invented has the effective power, area for area, that is inherent to the dipping lug in use almost invariably on those fishing craft which, by the nature of their work, have to make long journeys to sea and back again at their best speed. Here, instead of making many short tacks, only a few very long ones suffice to take the boat far to windward, and thus the delay caused by dipping is not appreciable. As a matter of fact, it is well known

motor became popular, the boats reverted to their old rig. I have had many opportunities of watching the two rigs in competition, and the difference was most marked. Indeed, it is not easy to describe the sense of irresistible yet easy power exhibited by one of these large luggers when

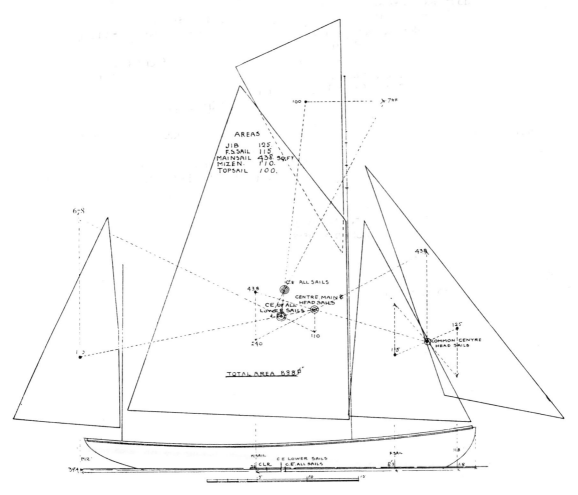

Fig. 37.

that for a year or two a good number of the larger type of herring boats, up to 60 ft. length, were tried with a fore and aft ketch rig of considerably larger area than could be obtained with the dipping big single lug and mizen; but the boats so rigged, though as fast on a broad reach, proved ridiculously inferior when competing with the lug-rigged boats close hauled. The innovation was soon given up, and, until the

sailing close-hauled in a strong wind. The tall, narrow sail, occupying the forward half of the vessel, seemed like some great seabird's wing cleaving the breeze without a tremor in its bronze curve, while the sharp bows chiselled their path with unfaltering speed through the sea.

Such exhilarating sights are, alas! now not so common, and it is a sad thought that in a very few years they will be seen no more; steam and oil are taking the place of sail so very rapidly.

Of course, such a rig is totally unsuitable for any kind of yacht, and the fact of its success in one type of fishing boat of comparatively light draught is only another proof that both boat and rig must be exactly adapted to the work required of them if perfect efficiency is expected.

If the advantages of the new cut of mainsail are desired in conjunction with a loose foot, there seems to be no obstacle to the combination if the work is done by an accomplished sailmaker. The method has been tried by enterprising sloop owners on the Humber, and favourably reported on.

The best material for all yacht canvas in sizes up to 40 tons is undoubtedly cotton. Egyptian cotton is more durable than even the best American, but the difference in this respect is not great. Cotton is, during its first season's use, especially liable to mildew, even when the greatest care is exercised, but, granting this drawback, its close, rigid texture, smooth surface and power of keeping its shape until the sail is worn out are advantages too great to be overlooked.

Flax canvas, on the other hand, is all that is undesirable for yacht work, especially in the lighter sizes. Loose in texture and elastic in substance, it seldom retains its shape for any length of time, though its wearing qualities, due in a measure to the nature of the material and its consequent lack of rigidity, are superior to cotton. It seldom mildews, and, even when badly worn, holds together to the very last. An old and worn cotton sail will go without any warning in the most unexpected way.

No cotton sail should ever be dressed or ochred until it has had at least one good season's use, and some of the starchiness washed out. If dressed on first bending it hardens and cracks across the threads.

The sail plan given in Fig. 37 is suitable for the hull shown in Fig. 25 (Part IV). It is treated diagramatically for the sake of clearness, in order to show the method of graphic determination of the different centres illustrated in Fig. 17 (Part III). For those who prefer the arithmetical method, which is not quite so expeditious, it will be seen that the various centres of jib, foresail, mainsail and mizen are dropped down to the waterline. Their areas are multiplied by their distances from a fixed point (in this instance, the stem), and the sum of the moments is divided by the sum of the areas, as follows:

	Areas.		Distance.		Moments.
Jib ...	125	×	1.8	=	225.0
Forestaysail	115	×	8.3	=	954.5
Mainsail ...	438	×	22.0	=	9636.0
Mizen ...	110	×	39.4	=	4434.0
	788				15249.5

$$\frac{15249.5}{788} = 19.35 \text{ ft.}$$

The distance of the Centre of Effort of *lower sails* abaft the stemhead, the topsail not being included, is 19.35 ft. This does not show the height of C.E. *above* the W.L. In order to obtain this the *heights* of the centres of the various sails above the W.L. are treated in exactly the same way. The graphic method, besides being quicker, gives the height as well as the distance aft of the stem in one operation, and in it the topsail is included and the height of C.E. for full sail shown. For ordinary comparisons the fore and aft distance of C.E. is essential. The height of C.E. is used in calculating the heeling moment under various wind pressures.

Many appliances are now in use for reducing expeditiously the area of head sails. Of these there are two very popular kinds. One is the "roller jib," which is a reefing and furling gear combined; the other is the Wykeham-Martin furling gear, on much the same principle as the roller gear, but without the rigid roller. It is a very effective means of furling the jib, and it obviates the necessity of putting the sail in stops previous to setting, thus making the process of shifting jibs in bad weather far easier than it used to be. It is applicable to almost any size of jib. Head sails intended to be used with this gear furl better and more evenly if the wire luff rope is parcelled out to a fair size. This assists the rolling very greatly. The Wykeham-Martin is not a *reefing* gear.

In the smaller sizes of cruisers, particularly the single-hander, where a single headsail is sufficient, the roller jib, if properly fitted and carefully used, is a great boon. Too often it is improperly fitted, and gives trouble in consequence. Fig. 38 shows a form which has proved very effective in many small craft. The points to be observed in the fitting are, in the first place, not to use it as a spinnaker. In the second place it is most essential that the roller be kept as straight as possible. There is, of course, always a tendency for it to curve when in use, owing to the wind pressure. To obtain as rigid a luff as possible the wire should pass through a hole in the bowsprit

the roller at short intervals. Strong twine should be used for this seizing. Before the fillet is put in its place the slot should be filled with tallow or lard. The tackle enables the luff to be kept taut. I prefer a lignum vitæ flange for the lower end of the roller in preference to a brass one. Brass and flexible wire do not agree, especially if wetted with salt water, and if brass is used the wire soon perishes. The flange rests on

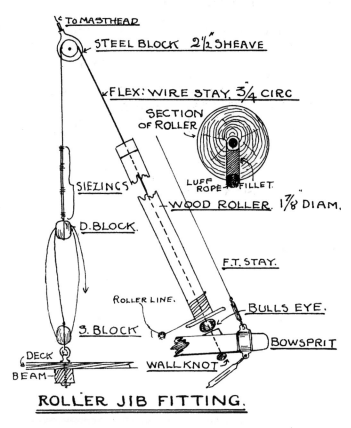

ROLLER JIB FITTING.

Fig. 38.

end, the wall knot (which is made solid by being dipped in solder after it has been made) preventing the end from coming through. The wire is laid in the slot in the roller, then passes through a light steel block at the masthead, and down to the upper block of a luff tackle purchase. It is not spliced to this block, but seized securely. The wire is kept in its place in the slot by a fillet of wood, which just fits in, but is not fastened at all, being kept in by the luff rope of the jib, which is seized to

a lignum vitæ bull's eye, to enable it to revolve easily. It is difficult to use satisfactorily a roller jib of much more than 120 sq. ft. area. If the roller is made of light wood free from knots, such as clean spruce or yellow pine, the extra weight forward is not very great. The convenience of being able to furl, or effectively shorten, the headsail from the cockpit is very great to a single hander, and a well-cut roller jib will stand excellently, however small the area exposed. But it is absolutely essential that the roller wire should be kept as taut as possible, and the foot of the jib kept at a flattish angle.

SPARS.

The best spars for small craft are those of Norway spruce, this being tough without undue weight. As little as possible of the outer skin of the spar should be removed in making. Spars, of course, are made of other woods, notably Oregon and pitch pine, the latter being heavy without any particular advantage beyond appearance. Oregon is very suitable for all spars when clean and of good quality. When varnished it is of a deep orange hue. Spruce, on the contrary, is pale yellow.

The diameters of spars vary, of course, with the amount of work they are called upon to do. A tall spar supporting a small and well-distributed sail area need not, if properly stayed, be of very great diameter. The usual proportion of thickness from deck to hounds is in craft of from three to five tons—.021 of the length. Thus a mast for a small yawl, 20 ft. from deck to hounds, would be .420 ft. in diameter, equal to a trifle over 5 ins. For vessels of from ten to twenty tons the multiplying fraction should be .023, and a mast 26 ft. deck to hounds would thus be .598 ft., or a little over 7 ins. If of very good quality the spar need not exceed 7 ins., and this diameter is sufficient for Oregon or pitch pine. Masts should taper but little from deck to hounds.

The thickness of a boom (for a laced sail) should be its length multiplied by .014. For an unlaced sail the thickness of the boom at its strongest part in the region of the sheet should be .016 of its length, gradually tapering towards the mast, but tapering more suddenly from the sheet to the outer end.

Gaffs have a diameter in the centre of .015 their length in small craft. In larger vessels this proportion should be increased to .016 or .017. They should be nicely tapered at the outer end, but not to too great a degree. A gaff has to be rigid if it is to do its work well, and with the present fashion of having the outer span at the extreme end of the gaff in order to keep the spar as much as is possible parallel to the boom, it is not wise to weaken this part of the spar too much, or it is apt to buckle up, and so put an undue stretch on the leach of the sail.

Bowsprits are now so much shorter than formerly, that their diameters need not be more than .037 of the length outside, and they may taper to .033 at the end. Oregon is a good wood for bowsprits.

Hollow spars are, of course, of greater proportional diameter, but the great saving of weight is an advantage in cruising as well as racing yachts. The manufacture of hollow spars is a very special industry, and is in the hands of only a few good makers. Spars of this kind are consequently very expensive. These spars are usually made of a very even grained wood, free from knots, called Canadian spruce. I have seen hollow spars made of bass wood, as well as yellow pine, which seemed quite equal to the work they were called upon to perform. No holes should be bored in hollow spars, and they should be solid at all parts where special local stress will occur. " Built " masts are made on the same principle as a good split cane fishing rod, and are of less proportional diameter, and therefore lighter than " grown " spars. They are also costly, and for cruising yachts quite unnecessary. The advantages of hollow gaffs, booms and yards, however, are very marked when used on cruising vessels.

The placing of the mast has, since the introduction of overhanging bows, undergone very considerable modification. In the days of the straight stem yacht the orthodox position of the mast of a cutter was about .4 of the length of the W.L. aft from the foreside of the stem, and the mast of a yawl from .35 to .38 of that length. This position of the mast was sound enough in principle, especially as the yachts of those days had very long and fine bows, incapable of supporting the weight of a heavy spar which was too far ahead of the centre of buoyancy. It had a grave disadvantage in respect to the sail plan in that it entailed a forestaysail of disproportionately large area. The foresail has ever had the reputation of being a very pressing sail, and it is the first sail to be lowered in a squall when of large area.

But to-day cruising cutters have the mast in a much more forward position relatively to the fore end of the waterline. In England a cruising cutter of moderate overhang forward may have her mast only .33 of the waterline length aft from its fore end, which may be considered a fairly satisfactory position. A yawl's mast may be .3 only of this length without adversely affecting her behaviour at sea. If lightly sparred, it may be even less.

Many fast fishing craft from the region of Yarmouth and Lowestoft carry their foremast about .22 of the W.L. length aft of a straight stem. These boats are ketch-rigged, with a boomless mainsail, and have the reputation of being fine sea boats. They are often sailed without bowsprit or jib in strong weather, which may be taken as an indication that the position of the foremast is quite as near the stem as is compatible with good behaviour. Had these boats a fair fore overhang this necessity would probably not arise, and they would be able to carry a small jib on a short bowsprit in any weather.

RIGGING.

For standing rigging plough steel wire is now invariably used in the smallest of cruising craft. It is unstretchable to any noticeable extent, and of enormous strength in relation to its weight and circumference. Properly prepared by being soaked in oil before it is shipped in place, and having all splices served and covered it is also extremely durable.

For running rigging in racing yachts flexible steel wire is very largely used, being set up by purchases. It also has the advantage of being practically unstretchable after a little use, and holding very little wind. But it is not nice to handle, unless one's hands are of the horny kind. Nor is it so durable as plough steel rigging wire. Friction through blocks and the wear and tear of belaying very soon open the strands and allow wet to get to the core. Rust then soon destroys it.

For the cruiser good manilla rope of fair size is preferable for all halyards. For the main sheet, Italian hemp; and for the jib and fore sheets the same, or, if preferred, cotton. Many people prefer three-strand rope to four-strand, on the ground that it stretches less and wears better. The heart of a four-strand rope often breaks and shows through the strands. There is very little difference, so little that it is immaterial which kind is used, provided it be of good quality. Rope of any sort but steel stretches and continues to stretch to the last, consequently in a long day's sail " setting up " will be necessary far more frequently than when wire is used for halyards. But rope has advantages over wire in being far more tractable to handle. It can be coiled, it will remain where it is put, and it can be cut with an ordinary knife in an emergency. On the whole the cruiser will be well advised to use rope for all running rigging.

In small craft the standing rigging may be set up quite satisfactorily by means of fine wire lanyards, which are neat and inexpensive, as well as strong and unstretchable. In larger craft rigging screws are essential, but these must be of the best pattern, and have locking screws. Small yachts with short bowsprits can dispense with bowsprit shrouds and a bobstay purchase, using in its place a strong rigging screw to set the bobstay taut. The bobstay should be of plough steel wire, and stouter than any other piece of rigging in the ship. The lower end should be secured to the eye on the stem by stout steel lugs and bolts or a strong shackle. It is a good plan to thoroughly grease the bobstay wire several times during the season.

Masthead fittings should be either light, well-finished wrought-iron bands, with the necessary eyes, carefully fitted to the taper of the masthead with only a very slight shoulder to support them; or the peak and main halyard blocks should be slung and supported by thumb cleats. This is the plan adopted by racing yachts, and it has the advantages of lightness and ease of removal. Lightness, so long as it is combined with strength, is as essential for the cruising yacht as the racer. Far too often the masthead ironwork seen on cruising yachts is needlessly heavy and preposterously strong, and far too often the masthead is weakened by the cutting of deep shoulders or the driving of heavy peak block and other bolts. In fact, there is less necessity for heavy ironwork aloft in the cruiser than there is in the racer. All the cruiser's gear may be at least as light as the racer's, provided it is well kept and properly examined from time to time. This cannot too often be insisted on, for one of the reasons why the performance of the cruiser under canvas is so often unsatisfactory is the clumsy, needless weight of spars, rigging and gear, and the indifferent cut, setting and care of the sails. These negligences often detract from the yacht's sailing ability to an astonishing degree.

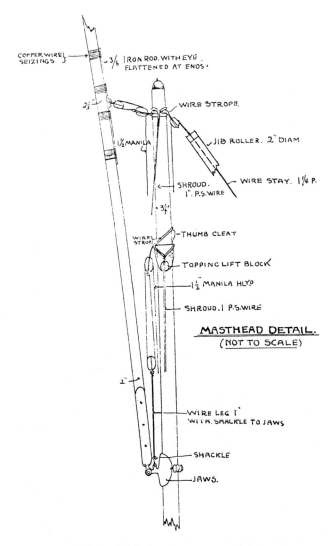

COPPER WIRE
SEIZINGS

~3/8 IRON ROD. WITH EYE
FLATTENED AT ENDS

WIRE STROPS.

2½"

1½ MANILA

JIB ROLLER. 2" DIAM

WIRE STAY. 1¼ P.

SHROUD.
1". P.S. WIRE

~3½"

WIRE
STROP.

THUMB CLEAT

TOPPING LIFT BLOCK

1½" MANILA HLY?

SHROUD. I P.S. WIRE

MASTHEAD DETAIL.
(NOT TO SCALE)

1"

WIRE LEG I
WITH SHACKLE TO JAWS

SHACKLE

JAWS.

A rigging detail taken from Strange's design for
DRYAD, a 23ft dayboat.

---

1 From page 78.

This section on the order of construction is of considerable interest. A well-fitted and well-sealed iron keel bolt can last 15 or 20 years before it needs replacing. Clearly, a better job can be made of it if the wood and iron keels are bolted together while access to both is easiest – before the remainder of the hull is built. The 'large stud bolts or coach screws', which then of necessity must be used to fasten the iron floors to the wood keel, are perhaps not initially so strong as through-bolts; but the crucial point is that they can be replaced without dropping the ballast keel. The average boat with through-bolted floors may have had its keel bolts checked and replaced assiduously over the years, but if the floor bolts have been neglected because of the major work involved in dropping the ballast keel, then the point will be reached where the only component holding the ballast/wood keel assembly to the rest of the hull is the garboard planks. This in one reason why an old hull frequently begins to leak first in this area. (See Fig. 31, page 68).

The Design and Construction of Small Cruising Yachts.

BY

ALBERT STRANGE.

VIII.—CABIN PLANS.

THE planning of the accommodation of a cruising yacht presents a good many problems that are very difficult to solve, especially in the smaller sizes. Though yachts vary in size, the human form is practically invariable in its requirements as to length, breadth and height, and though in many cases the demands of height have to be ignored, yet length and breadth must have their proper allowance if the yacht is to be a comfortable habitation, at least as far as sleeping accommodation is concerned. We therefore proceed by units of 6 ft. by 2 ft. in our " lay out " of cabins and forecastle, these dimensions being those of an ordinary comfortable sleeping space in our sea-going dwelling.

Height, of course, depends upon the size and draught of a yacht, as at least one or two berths six feet long and two feet wide can be obtained in the very smallest sizes of cabin cruisers, say, those about 18 ft. l.w.l. by 6 ft. beam. But in such small yachts the height of cabin cannot be much greater than about four feet, including the cabin top, if this is of the fixed kind, which will be found to be, on the whole, the best for general purposes. To obtain much more headroom than this will mean either that the yacht must have excessive freeboard and height of cabin top, or else that the draught and displacement shall be very large. Whichever way the extra height is obtained, it will be at the expense of the sailing qualities of the boat, as a high freeboard means a high centre of gravity, and a very deep body means a low centre of buoyancy, both of which, especially the former, tend to reduce the initial stability of the yacht, and thus detract from her sail carrying power and her general speed and handiness.

It is a fairly good, if quite approximate rule, that you can only obtain a trifle more headroom than your boat has draught of water if she is of a normal keel type, and if the cabin top is to be kept at a moderate height. If a good width of cabin sole is desired in the smaller sizes, this will help to take away from the height, and it will be for the owner or designer to choose between comfort for the feet and comfort for the head. A cabin sole of only 1 ft. 6 in. width is very inconvenient, except in a single-hander, where a companion is rarely carried, and it is difficult to extract much repose from a berth only 1 ft. 6 in. wide, more particularly in cold weather. It is a good practice to commence cabin planning after the method shown in Fig. 39, which takes account of the actual dimensions available. To start from the deck plan alone is very apt to lead to grave miscalculation as to the amount of room actually available.

The most important question in relation to height in the small sizes is the provision of *sitting* height under deck above the side seats or berths. This should not be less than 2 ft. 10 in. clear between cushions and carlines, and this amount is unobtainable in many small craft, rendering it impossible to get a comfortable seat in which one can recline at ease anywhere under the side decks. The only plan by which an easy seat can be obtained in very small boats is shown by the sketch, Fig. 40, where the

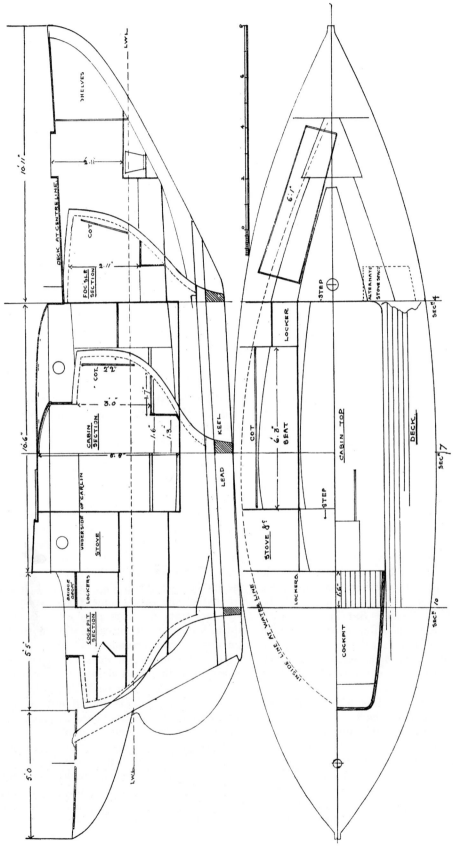

"LAY-OUT" OF CABIN PLAN FOR A 7-TON YAWL. L.O.A., 31.8; L.W.L., 24.0; Beam, 8.2.

Fig. 39.

side seats are carried across the boat some distance abaft the fore-end of cabin top, and a canvas seat, similar to the back of a deck chair, is attached to the deck beam and to the front of the 'thwartship seat. This plan is available in centreboard boats as well as keel boats, and as the canvas back is quickly detached, it will be found extremely convenient and comfortable.

The fitting of small shelves against the yacht's side under the deck (even where there is insufficient headroom between the cushions and the underside of the deck) is not to be recommended. Nothing more than a short piece of net-rack, as far as

floor to the top of the cushions. More than this verges on the uncomfortable, especially for ladies, and particularly if the seat is wider than 16 ins.

The plan usually adopted in small yachts of entering the cabin from the cockpit by means of doors opening directly into the cabin is one which not only entails a great waste of valuable space, but it is also one which deprives the yacht of a very essential amount of 'thwartship strength at a point where it is most required. The " bridge-deck " method of arrangement, or the provision of a wide thwart aft of the cabin, both of which give space underneath for

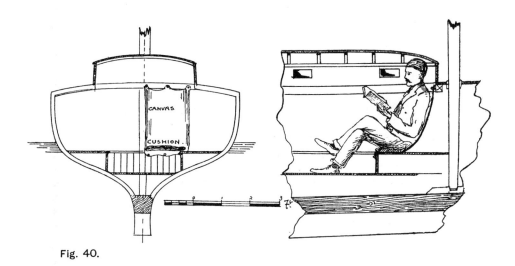

Fig. 40.

possible at the aft end of the space, is allowable if comfort is to be obtained when the crew is lying down. The practice of using the cushions for sleeping on is not a good one. The cushions become hard in a very short time, whatever may be their original composition, and far more repose is obtained from a cot berth which turns up against the side in the daytime and contains all the sleeping gear—blankets, pyjamas, etc., of the occupant. But cots should have a small hair mattress as a foundation for the bed, or they will be found very cold sleeping places. The height of the side seats and 'thwartship seats need not be more than from ten to twelve inches in small craft. In the very largest vessels the height of sofas and other seats is sufficient if it is not less than 15 ins. from the

lockers and cooking arrangements, and also very greatly increase the sitting accommodation in the cockpit, are by far the best afterend arrangements in yachts of from three to eight or ten tons. When cabin doors open into the cockpit so much of the cockpit side seats as represents the width of each door is rendered useless each side. If the doors open into the cabin, the same defect is felt, and although a little less facility of entrance and egress from the cabin is caused by the necessity of mounting over the bridge-deck or thwart each time, this inconvenience is less real than may be supposed, while the actual saving of valuable space is very great, and the cutting off of draught and the exclusion of rain from the cabin is also much assisted. The angle of bridge-deck and cabin top makes a very comfortable seat each side of the cockpit, and the actual length of cockpit may be much reduced without spoiling its comfort or effectiveness, by the adoption of this plan.

The question of the necessity of a watertight cockpit in small cruisers is one which has been debated for many years without any real decision having been formed. From an experience extending over thirty years I am inclined to draw the conclusion that, so far as safety is concerned, there is no need for a watertight cockpit. In a long, hard turn to windward there is no doubt that a good deal of spray does find its way aft, and if there is no watertight cockpit this spray deposits itself in the bilges. But it must be remembered that unless the floor of the watertight cockpit is very high—far above the level of the waterline—the heel and the speed will cause the outside water to back up the drain pipes and make the cockpit a very uncomfortable place, unless these drain pipes are plugged. The cockpit, although watertight, is not then self-emptying, and unless the accumulation of spray is got rid of in some way there is not only discomfort aft, but an accumulation of weight above the waterline just at a time and at a place where it is most objectionable.

It would therefore seem that in a small vessel, where the placing of the cockpit floor high enough to prevent this backing up is only obtained by making the cockpit so shallow as to afford too small an amount of shelter for comfort, a watertight cockpit is rather a nuisance than otherwise. The apprehension that a cockpit may be filled by a sea when sailing in broken water or in running before a breaking sea, is, of course, one which presents itself to the mind in contemplating extensive cruising. But it should be remembered that if the drain pipes have to be plugged no water can escape, and in the event of four or five hundredweight of water being thrown into the cockpit, this is retained there, and by being retained puts the yacht further down by the stern, reduces her lifting power aft, and really paves the way for a further invasion. It would seem better to allow the water to find its way into the main body of the boat, where it would be lower down and less dangerous, and if a semi-rotary pump of good size were fitted in the cockpit near

enough to be worked by the steersman, a lot of this water could be pumped out immediately. [1]

The fact that during a lengthy experience of cruising in small boats, I have never seen a sea break over the stern in a dangerous way does not, of course, dispose of the possibility. It *may* happen, but it is undoubtedly an extremely rare occurrence, and I think that if it did take place it would be better to have the water amongst the ballast than above the waterline. The weight saved in construction, and the large amount of stowage gained by the use of a cockpit whose floor will lift up, are considerations worthy of regard. The advantage of a watertight cockpit is more discernible in a very wet climate, when the boat may have to be left at moorings for some time. In this case no water ever gets below, and the interior of the boat is kept in a much dryer condition than it otherwise would be. In some wet seasons on the west coast of Scotland it is not uncommon for an inch of rain to fall during a night, and this may go on for several nights. If the boat is unattended it is surprising what a lot of water gets below unless kept out by means of a tent or a watertight cockpit.

Turning to somewhat larger sizes of yachts, the single cabin plan is found generally the most convenient in lengths of from 23 ft. to 26 ft. W.L. In boats of these lengths a lavatory is often desirable, and if there is a fair depth of body and a cabin top this is best arranged forward. A width of 2 ft. 3 in. and a height of 4 ft. 6 in. are about the smallest dimensions in which a satisfactory arrangement can be made. In small yachts on board of which no paid hand is carried, the affair may be placed quite in the eyes of the boat in the forecastle; one of the lightest types of closet being chosen, having the soil pipe carried up with a high curve before descending to the under water discharge. This arrangement gives good room, excellent ventilation, and a fairly satisfactory amount of privacy, the two latter conditions being the most essential, and, at the same time, the most difficult to secure in small yachts.

In a vessel of ten tons and upwards, the desire to obtain two cabins and a lavatory arises, but it is really far better to go to a size a few tons larger, unless the accommodation is to be of a very small and cramped

1 See page 96 for footnote.

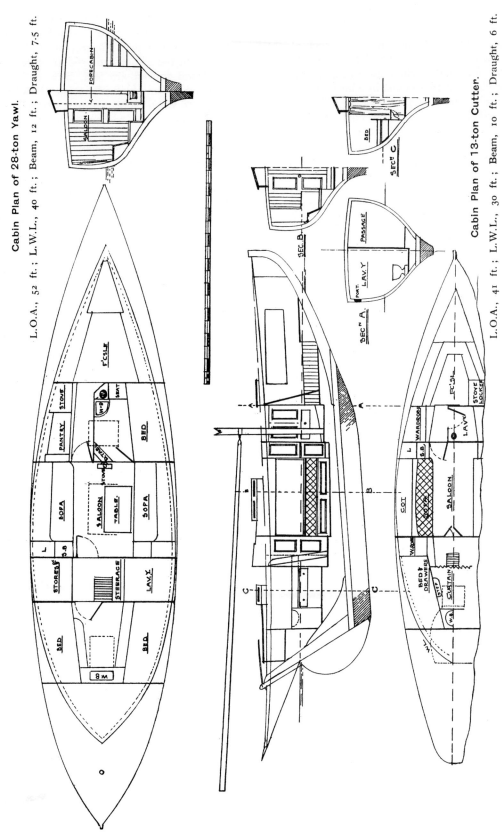

Cabin Plan of 28-ton Yawl.

L.O.A., 52 ft.; L.W.L., 40 ft.; Beam, 12 ft.; Draught, 7.5 ft.

Cabin Plan of 13-ton Cutter.

L.O.A., 41 ft.; L.W.L., 30 ft.; Beam, 10 ft.; Draught, 6 ft.

SKETCHES OF CABIN PLANS FOR CRUISING YACHTS.

nature. The difficulty is the case of a 10-ton yacht is often partially overcome by entering the sleeping cabin from the cockpit, and placing saloon and lavatory forward, aft of the forecastle. Such an arrangement is not well suited to a vessel in which winter cruising is contemplated, nor is it at all comfortable if ladies form part of the ship's company. Under the old 1730 rule, when yachts were long and narrow, two cabins and lavatory were more easy to plan, as a yacht 35 ft. L.W.L. by 8.5 beam and 6.5 draught allowed sufficient room. But in any yacht of modern proportions and of no more than 10 tons T.M. it can only be arranged by having part of the sleeping berths, some two feet or so, carried under the deck at the cockpit sides. This can be done if the yacht is full bodied and practically flush decked, but everything will be small and somewhat cramped. When the sleeping cabin is entered from the cockpit it is almost impossible for the inmates of the sleeping cabin to obtain an uninterrupted night's rest, especially on a night passage, as everybody must pass through to get on deck, and a good deal of rain or spray is likely to get below in bad weather. This arrangement is, however, very often adopted on "racing cruisers," even of 20 tons (as in the Clyde 20-ton One Design Class).[2] It is a more convenient plan for racing than for cruising requirements, the berths forming excellent receptacles for jibs, spinnakers and topsails, often very wet, but handy to get at.

It may be urged that the remedy for this is to put the sleeping cabin forward, with lavatory next to forecastle. This again has its serious drawbacks when a crew is carried. At meal times the crew has to come through the sleeping cabin to reach the saloon aft, or else bring the food over the deck. If ladies are on board this is a rather more unsuitable plan than any other, as a little thought will very soon discover.

The best arrangement by far, taking all things into consideration, is that which, whether in a flush deck yacht or in one having cabin top, allows entrance from a central ladder placed in a small steerage between sleeping cabin and saloon. With a cabin top this means a side companion (which should slide fore and aft, and *not* athwartships). By this lay-out complete privacy is obtained for the sleeping cabin

and lavatory, and entry for the purpose of setting table, etc., in the saloon is straight from the forecastle. Conversation is not heard from the saloon, and the arrangement holds good for all amateur crew or for a paid hand. The chief advantage, however, in the case when ladies are part of the company is that access to the lavatory can be had without passing through a cabin for the purpose, and this, especially at night, is well worth securing.

For yachts of small tonnage, and those which are used for single-hand cruising, there are other arrangements that may be preferred. The central position of the average swing table is often not the most convenient, especially in a cabin having less than 3 ft. width between the lockers athwartships. Where there is only a crew of one or two, a good arrangement is to have the table hinged to either side of the yacht, or to the forward bulkhead. It can, of course, be made to unship and stow away in very small craft. But it will generally be found that any fitting that has to be fixed up and put away after each meal soon becomes a nuisance, and is generally given up after a very short trial. Indeed, in small yachts the swing table is a somewhat doubtful blessing, whether of the ordinary type or slung from the cabin top beams. The movement of a small yacht is too quick and irregular to allow satisfactory balance when under canvas in rough water, and most meals taken when under way are of the picnic order.

The position of the galley, or cooking place, is one requiring a good deal of consideration, and depends very largely upon whether the crew is completely amateur, or whether a paid hand is carried to do this work and its attendant washing up. When the cooking is performed by the paid hand, it is, on the whole, far better to provide a special small galley in the forecastle, having drawers and lockers for the stowage of utensils as part of its design. The somewhat rough and ready methods of the paid hand are then not quite so obvious, and much passes unseen that might otherwise take the edge off appetite. But it is important that a paid hand should not be expected to do the cooking *aft* of the saloon, or he will be there on every possible occasion, and his society will frequently be found *de trop*.[3]

See page 96 for footnotes 2 & 3.

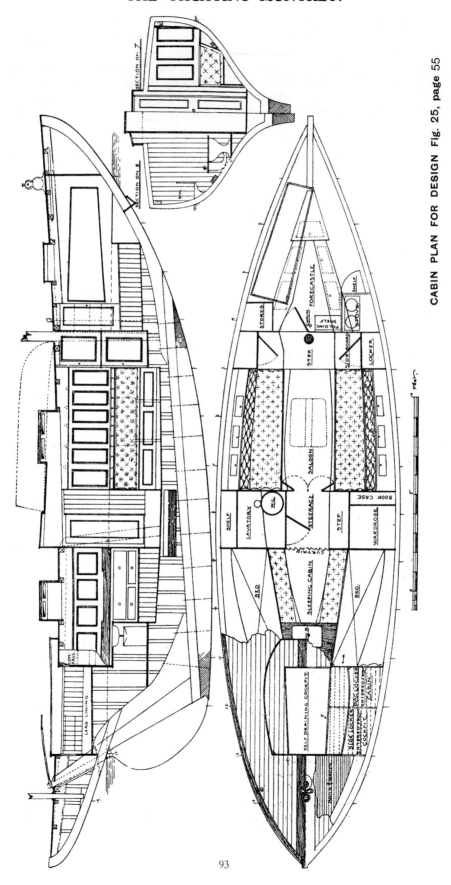

CABIN PLAN FOR DESIGN Fig. 25, page 55

But if no paid hand is carried there is much to be said for the aft position of the galley, as shown in the cabin plans in this article. There is much more headroom, ventilation and convenience of distribution after the food is cooked, and a small sink can often be arranged on the opposite side, some few inches above the waterline for washing up. The sink should be shallow, and have a screw plug to the discharge pipe. A properly gimballed double primus stove can be arranged in a fore and aft length of a trifle over two feet, care being taken to allow sufficient room under the stove to allow it to swing unchecked. But, on the whole, a stove lashed in place will be found to give satisfaction, especially in these days of Thermos flasks. Cooking under way requires a good deal of attention and intelligent apprehension if accidents of a most disquieting nature are to be avoided, though the short operations attendant on the production of hot soup or a cup of tea can generally be carried out even in bad weather if a member of the crew is free to attend solely to the business.

Locker space is not difficult to arrange, especially with the bridge deck, as this affords a large clothes locker (an unusually valuable adjunct) in the centre, and four other store lockers each side and under the deck. The usual sideboard and locker forward of the saloon sofa can generally be adopted, as in this part of the boat the space that they take up is not very easily utilized for other purposes. These lockers and sideboards need not be more than a foot wide to afford a lot of storage. Lockers with horizontal or lifting lids may be considerably less than a foot wide, and yet contain a good deal of room easily got at.

All lockers should have small ($\frac{1}{2}$ in.) holes bored in their sides or bottoms, above and below. A row of about eight will be found to give good ventilation. The lockers should be varnished inside. They can then be washed out and dried easily, a quite necessary operation, seeing how many things get capsized and spilled in the lockers during a short cruise.

Uncooked meat is often extremely difficult to keep on any yacht. A small ventilated meat safe with a zinc tray is one of the best things for keeping meat in. The safe should be on deck in an airy place. It is not generally known that dilute ammonia, dabbed on the inside of the safe, and on the tray, is of great service in preventing meat from " turning " in hot weather. It also keeps flies off.

The fresh water tank will generally be found to stow well under the steerage sole. It should be connected to a small rotary or " beer-engine " pump. In all cases it is best to have the water tank specially made to fit its position. The bottom should slope, and at its lower end should be fitted a tap, which enables the flushing and cleansing water to be run off. The hole in the top of the tank should not be less than eight or ten inches in diameter. The frequent cleansing of water tanks is very necessary, but very few builders, if left to themselves, will take the trouble to provide taps or sufficiently large hand holes. Each tank should, of course, have a filling pipe from the deck, and an air pipe, with the open end turned over to prevent dust or dirt from entering. An allowance of one gallon per unit of crew per day is enough.

Light is obtained either by the use of port lights in the coamings of the cabin top, or by a skylight when yachts are flush decked, or having what is called a companion skylight. This arrangement gives headroom without so much excess of weight as is the case with a cabin top, but is most useful in yachts which have a good depth of body. It is well to avoid too round a curve in the coamings of a cabin top; the excessive round renders the stowage of spars, oars and boathook very difficult. These impedimenta lie very snugly alongside a companion skylight.

Yachts fitted with a cabin top and the consequent narrow side decks, should have a hand rail fixed on the roof of cabin top. The best height for this rail is about 3 in., and it should be placed about a foot inside the outer edge.

When a yacht is riding head to wind the current of air through the cabin is always from aft. Complete ventilation can be obtained at night by the adoption of louvres in the cabin doors or slide, and by raising the fore hatch an inch or so on wedges. If the slide is in two pieces the top one can be left out, or the companion slide pushed back an inch or two.

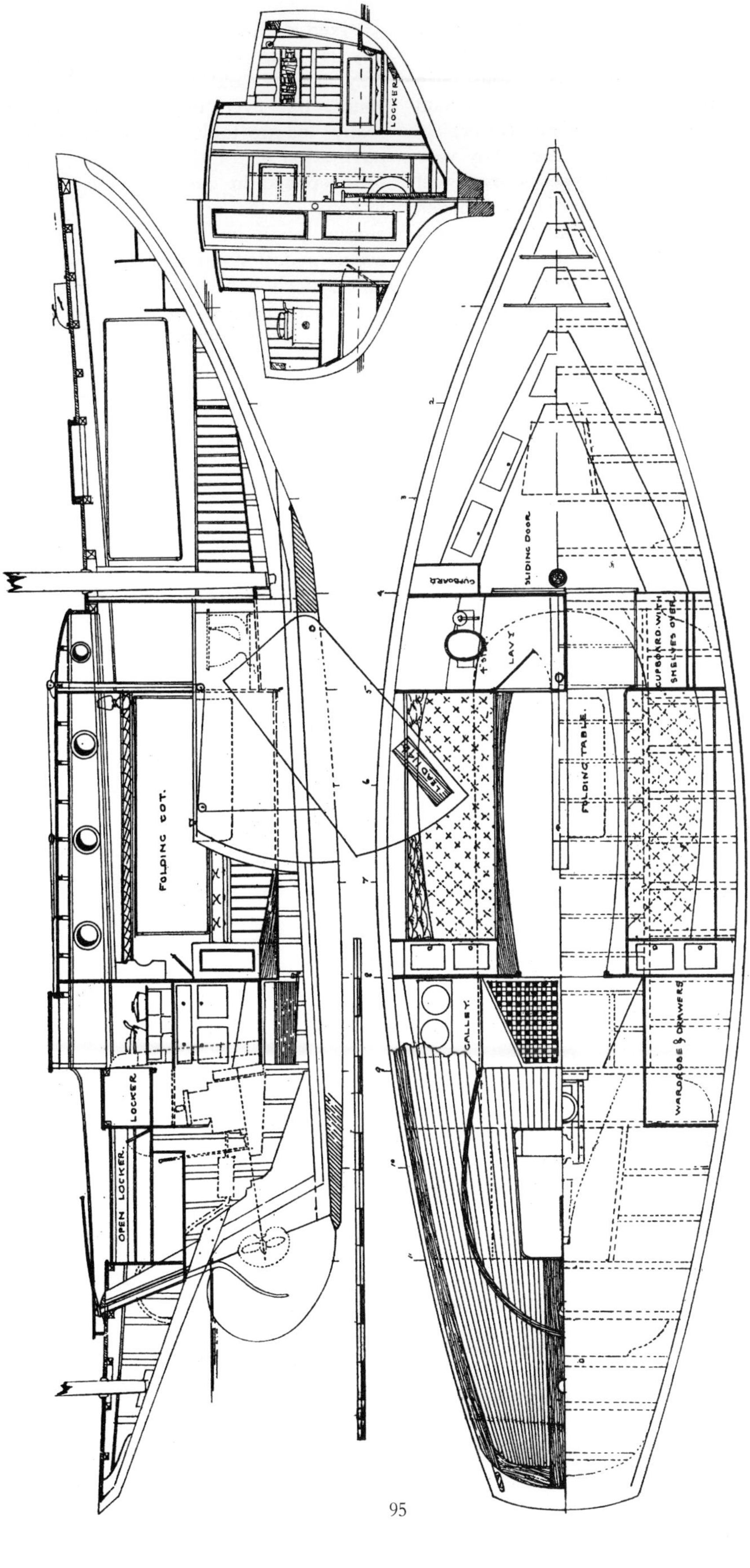

LOCKER.

FOLDING COT.

FOLDING COT.

OPEN LOCKER. LOCKER.

SLIDING Door.

CUPBOARD.

TOILET

LAV!

FOLDING TABLE

CUPBOARD WITH SHELVES OVER.

GALLEY

WARDROBE & DRAWERS

CABIN PLAN OF 25-FT. WATERLINE CENTREBOARD YACHT.   Fig. 28, page 62

In cabin planning there is much room for individual treatment, but it must not be forgotten that the arrangements shown in the illustrations are those which commend themselves to the general cruising fraternity, having been proved by long use. Any very great changes from the types shown will not only have their own special drawbacks (as have all compromises), but will perhaps not recommend themselves to anyone but the person who planned them—a fact not to be lost sight of when market requirements are considered.

Fixed double berths are not desirable in any cruising yachts under 60 tons. Not only are they extravagant in the matter of space, but even in harbour they are not really comfortable if responsibility for navigation or for the usual care that should be exercised in berthing, etc., is assumed by one of the occupants. A double and single berth in the same cabin means accommodation only for two persons, unless a child occupy the single berth. In any craft with less than 14 ft.[4] double berths are very undesirable.

---

1 Strange's comments on the self-draining cockpit are worthy of careful consideration in these days when its presence is taken for granted and its drawbacks, some of them potentially dangerous, go unquestioned.

2 This class, designed by Alfred Mylne in 1899, measured LOA 51ft, LWL 35ft, Beam 11ft, Draught 7ft 6ins, SA 1700sq.ft. *Tigris* is a recently restored example. The lines, accommodation, and sail plans are to be found in Dixon Kemp.

3 The particular problems of accommodating both ladies and the paid hand in a small yacht have little relevance today, but the elegant language with which Strange addresses them is delightful.

4 Insert 'beam'.

The Design and Construction of Small Cruising Yachts.

BY

ALBERT STRANGE.

IX.—CONCLUSION.

THERE are a few points of great importance that have been left over for later attention, as they have occurred in previous articles, mainly for the reason that their investigation would have taken too much space at the time of their occurrence, and would have interrupted the consideration of the more elementary matters then under discussion.

One of these points of importance is the study of what takes place when a vessel is heeled from her upright position, and so undergoes a change of immersed hull form. We know already that when in an upright position the centres of buoyancy and of gravity are in the same vertical line, and that upon any alteration of trim, longitudinally or transversely, a struggle is set up and an effort to adjust themselves to their normal positions takes place, and continues until the vessel is once more in an upright position. This can only happen when the force or the weight causing the inclination is removed.

We can only study the change of form and the consequent change of the centre of buoyancy (for the position of the centre of gravity never changes relatively to the vessel under normal conditions) by adopting a highly artificial line of procedure. That is, to assume that the alteration of trim on heeling takes place unaccompanied by any progressive or forward motion of the vessel. In fact, we proceed on the assumption that the vessel is heeled in perfectly still water, such as might occur in a dock when the vessel would be moored fore and aft and heeled by ropes from the masthead. This is necessary because no accurate

diagram can be made of the immersed portions of the vessel when she is in motion.

But our diagram of the heeled vessel contains much information that bears vitally upon the vessel's performance under canvas and when in motion. The point of greatest importance is to discover what change takes place in the fore and aft position of the C.B. when the vessel is heeled. It is clear that a change occurs laterally by a movement of the C.B. towards the lee side of the vessel, but it is not always apprehended how great a change may take place in a fore and aft direction.

If we take a plan of a small yacht, as shown in Fig. 41, we can draw lines through the intersection of the L.W.L. and the centre line at any required angle of heel (in this case 24 degs.), as shown by lines A B Y Z passing through O, which cut the sections in C D E F, etc. These lines represent an *inclined* load waterline for both sides—that is why they are carried through to both sides. The development of the inclined load waterplane is shown below the half-breadth plan, and this is its actual shape *in still water*. It will be seen that it is of unequal, but fair, outline, and that the "immersed" side shown above the centre-line is much larger than the "emerged" side shown below. To find the points through which the surrounding curve must pass, measure from O to C on Sec. 2 and put the distance above, and then from O to D on the same section, putting the distance below the centre line. Proceed in like manner with each section, and draw the curve in by freehand, as some fairing up may be at first necessary. To find its termination aft measure up from the L.W.L. in body plan to the point J (on body plan), which is

97

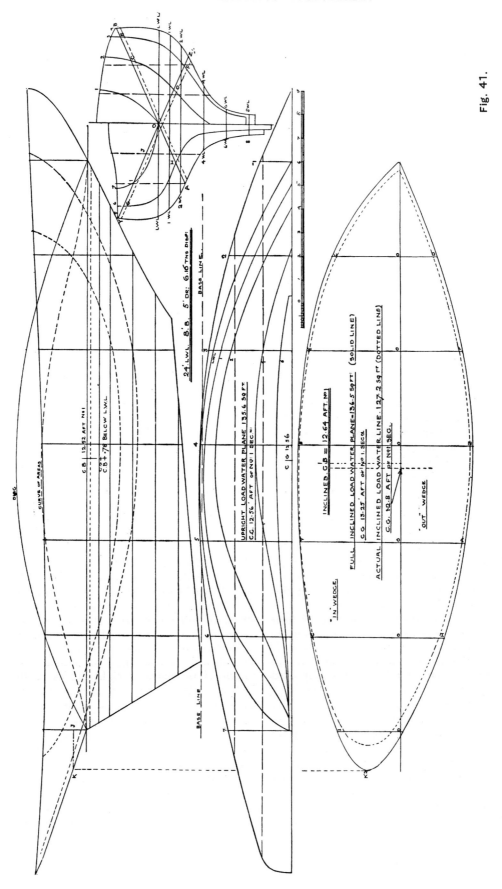

Fig. 41.

where the inclined line cuts the inner buttock line; transfer this distance up to J on section 7 on profile and square along horizontally until the line of counter is cut at K. Then square down to K2, which point will be the after end of the inclined load water plane. As the fore overhang is short, and not immersed on inclination, the forward termination of the inclined waterline must be at Sec. I.

It will be seen that the shape of the inclined load water plane differs very considerably from that of the load water plane in an upright position, and this difference is indicative of what may have taken place in the whole underwater body. If the position of the centre of gravity of this inclined plane is ascertained (as explained on p. 414, No. 102)[1] it will be found to be further aft than the same point on the load waterplane. This movement aft, though inconsiderable, would tend to depress the yacht by the head and lift the stern slightly if the C.B. of the immersed inclined portion of the whole underwater body acted through it. This, however, is not what happens. The C.B. *of the whole underwater body* must be found, and it will be through this point that the lifting tendency longitudinally would act at whatever place it was situated. This lifting tendency would, of course, help to depress the yacht's head if it occurred aft of the centre of gravity. As it succeeded in depressing the vessel's head it would tend to advance forward until checked by the resistance of the spreading shapes of the bow section and the piling up of the lee bow wave if the vessel were in motion, and, in moving forward, would eventually come to rest near the centre of gravity of the vessel, and remain there until the heeling force was removed.

It is sometimes argued that the centres of gravity of the two unequal divisions of the inclined waterline act in some separate way. But as the water takes no notice of an arbitrary "centre line," which is only used as a geometrical convenience, the centre of gravity of the whole inclined plane is the important point. But this point is not so important as the position of the C.B. of the whole inclined *underwater body* of the yacht, and this must now be ascertained.

The beginner will do well to draw out the sections as shown in Fig. 42—that is, the full sections. The upright displacement is first ascertained (in this case 215.6 cubic feet), then the inclined lines are drawn across through the centres of each as shown, and the inclined displacement measured—an easy matter if the designer possesses a planimeter, but more complicated if it has to be done by measuring the area of the immersed wedge and the emerged wedge, or "in" and "out" wedges, and subtracting the "out" wedge from the area of the "in" wedge in each section and adding the difference to the upright area of each. When this is done it will be found that the displacement of the *inclined* hull exceeds that of the *upright* hull by no less than 22.4 cubic feet, owing to the fact that in heeling more has been put in on the lee side than is taken out on the weather side.

Now we know that unless weights are added, the displacement of a yacht remains the same, plus the very small depression due to the downward component of the wind force acting on the sails. So the yacht, in order to accommodate herself to this increase on the lee side, must rise bodily until the inclined displacement equals the upright displacement. To find out how much the yacht will rise to do this we divide the excess displacement in cubic feet by the area of the *inclined* load water plane, as follows:

$$\frac{\text{Excess disp:} \quad 22.4 \text{ cu. ft.}}{\text{Area inc. plane: } 136.5 \text{ sq. ft.}} = .165 \text{ ft., or 2 in.}$$

We then draw a line parallel to the first drawn inclined W.L. at a distance of .165 ft. below (as shown by the dotted line on each section) and these are then the true lines, by means of which we can construct our inclined load waterline of the proper area. This is shown by the dotted line inside the solid line of the inclined plane, Fig. 41, and is the true inclined load water plane. If no planimeter is available, the areas of the inclined sections must be calculated by means of ordinates at right angles to the inclined waterline shown on Sec. 4, Fig. 42, or, better still, by the use of squared paper on which the sections must be drawn. The shape of the heeled underwater body is so irregular that neither Simpson's nor the trapezoidal rule will give the exact areas. These are very important, and must be exact if the calculation is to be of use. These areas, measured by planimeter, are shown on the shaded part of the sections,

1 Our page 47.

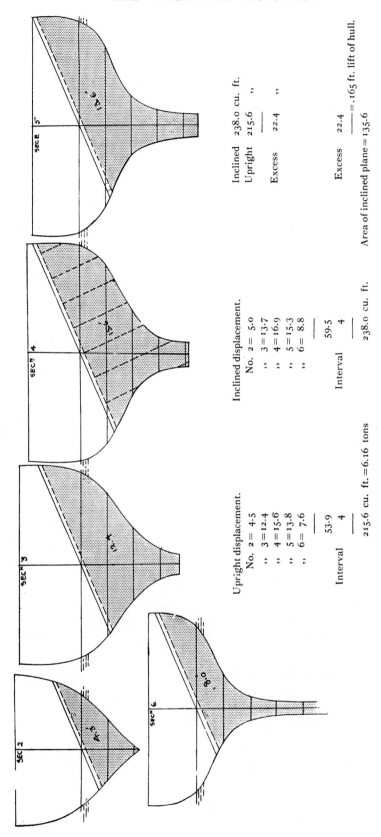

Upright displacement.

No. 2 = 4.5
,,   3 = 12.4
,,   4 = 15.6
,,   5 = 13.8
,,   6 = 7.6
          ——
          53.9
Interval   4
          ——
215.6 cu. ft. = 6.16 tons

Inclined displacement.

No. 2 = 5.0
,,   3 = 13.7
,,   4 = 16.9
,,   5 = 15.3
,,   6 = 8.8
          ——
          59.5
Interval   4
          ——
238.0 cu. ft.

Inclined 238.0 cu. ft. ,,
Upright   215.6   ,,
          ——
Excess   22.4   ,,

Excess   22.4
          —————— = .165 ft. lift of hull.
Area of inclined plane = 135.6

NOTE.—The "inclined displacement" is the first, or *uncorrected*, inclined displacement. The figures on the shaded part of the section are those of the "corrected" inclined displacement.

Fig. 42.

and can now be summed by the usual method and the true *longitudinal* position[2] of the inclined immersed portion of the hull ascertained. The whole working is as follows :

" Corrected " Inclined Displacement.

Moments
for C.B.

Section 1 = 0.0 × 0 = 0.0
,, 2 = 4.3 × 1 = 4.3
,, 3 = 12.4 × 2 = 24.8
,, 4 = 15.6 × 3 = 46.8
,, 5 = 13.6 × 4 = 54.4
,, 6 = 8.0 × 5 = 40.0
,, 7 = .1 × 6 = .6

Sum   54.0   170.9

Dis. of sec. 4 ——— = 3.16
            54

4 = Dis. of sections.

Cu. ft. = 216.0 = Disp.

12.64 = Distance of C.B. of inclined hull abaft No. 1 section.

It will be noted that even planimeter readings do not give a displacement that quite coincides with the upright, but the difference of .4 cu. ft. is too slight to be of consequence.

Thus we see that the position of the C.B. of the inclined immersed body is slightly aft of the position of C.B. when the yacht is upright, by the amount of .12 ft., a quite negligible distance. We may conclude then that the yacht would only alter her fore and aft trim to an extremely small extent when heeled if the heeling takes place in still water.

It must also be remembered that we now only know the *distance aft,* but not the true position laterally of the heeled C.B., which is necessary to complete the calculation. The best way to find out how far the C.B. has shifted to leeward is by the graphic method, done as follows. Cut out of cardboard the exact shapes of the true inclined *under water sections* as shown shaded on Fig. 42. Stick these one on top of the other in their exact numerical and relative positions, with the centre line coinciding exactly in each section. Thin paste may be used for this. When dry, balance the combined mass of sections on a needle point, or find the point of intersection of two or more suspensions, and where the point of balance is, there is also the exact position laterally of the in-

clined C.B. for the given angle of 24° only. If we could now find the position of the C.G. of the hull we should be able to ascertain the length of the righting lever at this angle. As before explained, this is a very long process, in which every part of the hull is measured and its weight and C.G. ascertained. For very particular work, such as for a motor boat, a light displacement auxiliary yacht, or a steam yacht, it is absolutely necessary, but for a keel-ballasted sailing yacht it is really less important. A fair estimate may be made as follows, but unless it is accurately and thoroughly done it may be dangerously misleading. The weights, multiplied by the distance of the C.G. of all parts, such as keel, hull, frames,

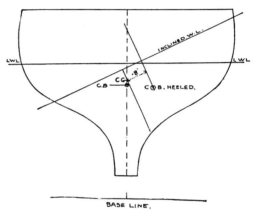

Fig. 43.

beams, deck, sails and spars, etc., are set off from a given base line, say, 5 ft. below the L.W.L. and parallel to it as follows :

Approximate estimate of position of C.G. of hull.

Weights (tons)		C.G. distance up from base line.	Moments.
Lead keel	2.5 ×	1.6 ft. =	4.0
Hull & deadw'd	1.8 ×	4.0 =	7.2
Deck & house	.8 ×	7.5 =	6.0
Beams	.15 ×	7.0 =	1.05
Mast, sails & rig	.30 ×	18.0 =	5.40
Ironwork	.10 ×	7.0 =	.70
Chain & anchor	.15 ×	3.0 =	.45
Stores, etc.	.30 ×	4.5 =	1.35

6.10 = Tns. dispt.   26.15

——— = 4.28 ft.
6.10

(4.28 is the distance of the C.G. of hull above the base line drawn 5 ft. below L.W.L.)

2 Should read ...position *of the C.B* of the inclined...

This position of the C.G. is, as stated, approximate, but it is not far from the true position, and the error, if any, is on the side of safety. A good deal of experience of construction, and plenty of data from which to make estimates, are necessary to obtain a near approximation, and the method explained should be used with great

We can now see how much sail it would take to heel the boat to this angle when the wind pressure is 1 lb. per square foot. Fig. 44 shows the sail plan and the centres of effort and lateral resistance.

The distance of the C.E. of sails is 18 ft. above the C.L.R. The sail area is 600 sq. ft. Sail area multiplied by 18 ft. $\times$ .1 [3]

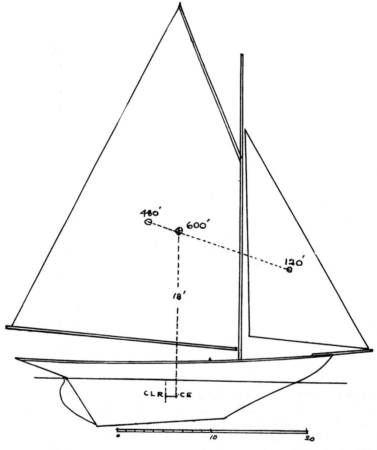

Fig. 44.

caution by beginners. It will be noticed that full allowance of weight for mast, sails and rig has been given, as this factor is of great moment in fixing the true position of the C.G. Fig. 43 shows the midship section with the true positions of C.B. upright and inclined, and also the position of C.G. The distance between the inclined C.B. and the C.G. gives the length of righting lever, .8ft., and the height of metacentre for this angle of heel is 1.85 ft. above C.G.

gives a heeling moment of 10800. This is divided by the displacement in pounds (6.1 tons $\times$ 2240 = 13644 lbs.)

Sail moments.

$$\frac{10800}{13664} = .797 \text{ ft.} = \text{necessary righting lever at } 24°$$

The righting lever of .8 ft. is therefore slightly more than the amount required. A wind pressure of 1 lb. per sq. ft. amounts

3 This is a mis-print; 18ft x .1 should be 18ft x 1.

to a smart breeze of about thirteen[4] miles per hour, or Force 5, and when the yacht has heeled to 24° it is about time to think of shortening sail.

The example given for working out these various calculations was chosen because it illustrates the necessity of careful balance of forms at the ends of the yacht if the trim is to be preserved at various angles of heel, and also to show what an astonishingly big lift takes place when a strong bilged model is heeled; apart from its being a very useful size of boat as an exercise in design.

When all the calculations have been made and the various centres ascertained, then remember, please, that they have all been considered under an almost impossible set of conditions. Under canvas the form of the waterline would be something widely different from the level lines employed, and the various centres would have shifted to an unknown extent. But it is quite certain from long experience that a design which will bear this geometrical analysis is likely to be a good and trustworthy performer at sea, and that any design whose centre of buoyancy shifts considerably forward or aft from its calculated position when upright will develop undesirable and highly astonishing qualities when sailing at speed, even in smooth water.

Area of wetted surface.—As will be readily imagined, this bears a close relation to the shape of the longitudinal immersed section. A "cut-away" underwater plane has naturally a reduced area of surface, but much pared down lateral planes are not good for cruising yachts, and any leanings towards an unusually small amount of wetted surface should be indulged in with caution. In the matter of speed, of course every foot tells, but it has been proved many times over that an increase of wetted surface adopted to cure erratic steering caused by a too short lateral plane, has resulted in greatly increased speed. At six knots the friction, or the "resistance," amounts to as much as $\frac{1}{4}$ lb. per sq. ft. of surface, when that surface is smooth. It may be doubled or trebled if the surface is rough with barnacles or weed growth.

The exact area of wetted surface is difficult to calculate, owing to the varieties of curvature in the underwater form of a yacht. The simplest of many methods is here given. The length of L.W.L. is measured

round its curve. In this instance the length of 24 ft. increases by curvature to 26.4 ft. This distance is divided into six equal parts,

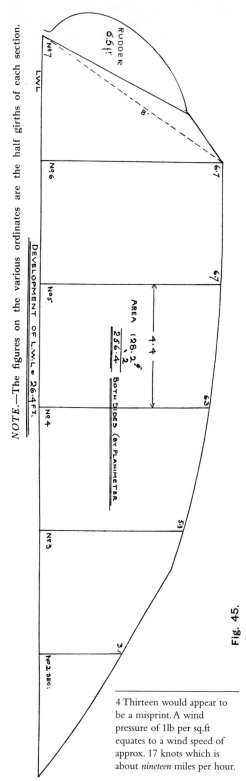

NOTE.—The figures on the various ordinates are the half girths of each section.

Fig. 45.

DEVELOPMENT OF L.W.L = 26.4 FT.

RUDDER 6.5'

AREA 128.2'

$$\frac{256.4}{4.2}$$

BOTH SIDES (BY PLANIMETER)

---

4 Thirteen would appear to be a misprint. A wind pressure of 1lb per sq.ft equates to a wind speed of approx. 17 knots which is about *nineteen* miles per hour.

to correspond with the number of sections. The half girth of each section is measured, and an ordinate corresponding to the girth length is set up at each division. These may be summed either by the trapezoidal or Simpson's rule, and the odd shape aft measured by triangulation, and added. The area of rudder also should be included. The result obtained by this method is slightly less than the actual amount.

In the example shown in Fig. 45 the area amounts to 182.2 sq. ft. for the hull and 6.5 ft. for the rudder, which should always be included. Then :

$$
\begin{array}{l}
\text{sq. ft.} \\
182.2 \\
6.5 \\
\hline
188.7 \\
2 \text{ for both sides} \\
\hline
377.4
\end{array}
$$

This area multiplied by .25 gives a resistance of 94.35 lbs. at six knots speed.

This resistance is due solely to friction, and does not include the greater resistance due to form when a high rate of speed is set up. The resistance due to form is not calculable, but can be obtained by experiment when towing. The resistance ascertained by this form of experiment is, of course, the total resistance, and that part of it due to form can be ascertained by deducting the amount caused by surface friction; obtained by the calculation just explained. At low speeds, say, from $2\frac{1}{2}$ to 3 knots, the resistance due to friction is very much less, probably not more than .1 lb.

per sq. ft., and the resistance due to form is very small in a yacht of 24 ft. W.L. at these low speeds. For lengths of from 30 to 50 ft., resistance, owing to finer lines and the fact that the particles of surrounding water are set in motion corresponding to the forward motion of the yacht, is even less.

It is hoped that this series of articles will have been of use to the amateur designer, and will encourage him to proceed further in the study of the fascinating subject of naval architecture. In preparing them I have endeavoured to simplify all calculations as much as possible, remembering my own early and unaided efforts to overcome the mathematical difficulties that seem so overwhelming when first encountered in more advanced treatises. When I think of the immense strides that have been made by the amateur designer of late years, due almost solely to the encouragement offered with persistent generosity by this magazine, I feel sure that some others of its readers will be assisted to further increase the number of those who desire to design their own boats. Few things are more fascinating and inspiring than to see the gradual growth of one's own ship, from the drawing board to the final and crowning glory of beholding her move forward on her path through the seas with power and certitude.

We, who were born on earth and live by air,
  Make this thing pass across the fatal floor,
The speechless sea : alone we commune there,
  Jesting with death, that ever open door.
Sun, moon and stars are signs by which we drive
  This wind-blown fabric like a thing alive.
                                            —*John Masefield.*

SCANTLINGS AND MATERIALS.

In order to assist the beginner in his estimate and calculation of the weights of the materials used in the construction of the hull, a table of scantlings is now given. These scantlings are not racing scantlings, but they do not pretend to be in all ways so substantial as those prescribed by Lloyd's Society for cruising yachts. With first-class material and workmanship they will, however, be found amply heavy and sufficiently strong for all cruising purposes, collisions with other vessels, quay walls, or floating derelicts excluded.

Table of Scantlings

L.W.L.	Approx Displmt.	Keel Siding	Stem & Sternpt. Sid. Mld.	Steamed Tbrs. Sid Mld.	Spacing Centres	Grown Frames	Spacg. Crs.	Plank. finished	Deck Plank	Shelf	Bilge Stringer	Main Beams Centres	Other Beams Centres	Spacing Centres
20 ft.	3½ tons	3½"	3¼" × 4"	1¼" × 1"	6"	-	-	¾"	1"	3½" × 1⅛"	2½" × 1⅜"	2½" × 1¾"	1½" × 1⅛"	9"
25 "	6 "	4"	3½" × 4½"	1½" × 1¼"	7"	Sid. – Mld.	-	⅞"	1⅛"	4½" × 1⅜"	3¼" × 1⅜"	3¼" × 2¼"	2" × 1¼"	11"
30 "	11½ "	5"	4¼" × 5½"	1¾" × 1⅜"	2 each Bay	2" dbl. × 2¾"	30"	1⅛"	1¼"	5½" × 2"	4½" × 1⅝"	3¾" × 2⅝"	2½" × 1⅝"	13"
35 "	18 "	6"	5" × 6½"	2" × 1⅝"	do.	2¼" × 3¼"	24"	1¼"	1⅜"	6½" × 2¼"	5" × 2"	4¼" × 2¾"	2¾" × 1¾"	14"
40 "	26 "	7½	5¼" × 7½"	2¼" × 1¾"	1 each Bay	2½" × 4"	20"	1⅜"	1½"	7" × 2½"	6½" × 2¼"	4⅞" × 3"	3¼" × 2¼"	16"
		moulding as required				moulding reduct. at heads		for p. pine, larch, Kauri p.						

Table of Scantlings and Sizes of Iron Floors, Knees and Fastenings suitable for yachts of from 20 to 40 feet waterline length.

The table is based on general practice at the present time.

Sizes of Iron Floors, Knees & Fastenings

L.W.L.	W.I. Floors Throats Points	Length of Arms	Hanging Knees	No. aside	Oak Lodg-ing Knees	No. aside	Deadwood Bolts	Knee Shelf-bts.	Plank fastening Steamed frames	Plank fastening Grown frames	
20 ft.	1¼" × ½" to 3/16"	14"	12" × ¾" × 5/16"	4	1⅛" thick	4	7/16"	3/16"	3/32"	-	One W.I. Floor on every other timber
25 "	1½" × ⅝" " ¼"	16"	13" × ⅞" × ⅜"	5	1⅜" "	5	8/16"	4/16"	⅛"	-	do. do.
30 "	2¼" × ⅞" " 5/16"	18"	16" × 1⅛" × ⅝"	7	1⅞" "	6	9/16"	5/16"	3/16"	¼" to 5/16"	One Iron Floor on each grown frame.
35 "	2½" × 1" " ⅜"	22"	17 × 1¼" × ¾"	7	2¼" "	7	10/16"	6/16"	¼"	5/16"	do. do.
40 "	3½" × 1¼" " ½"	28"	20 × 1¾" × ⅞"	8	2½" "	8	12/16"	8/16"	5/16"	⅜"	do. do.
			Tapered at each point								

NOTES ON MATERIALS.

Unless the purchaser is well versed in the knowledge of timber (a knowledge not easy to obtain except by the painful processes of trial and error), he will be well advised to employ an experienced person to buy for him, and to be prepared to pay a good price for the stuff. Do not be deceived into thinking that you can buy good timber at less than the market price. It is almost impossible to do this in ordinary circumstances. The following woods are those chiefly used in yacht building.

British Woods.—Oak, elm, larch, and sometimes ash and fir.

Foreign Woods.—Teak, American or Canadian elm, white oak, pitch-pine, Kauri pine, yellow pine, mahogany, Oregon pine, cedar.

Other timbers are occasionally used, but these are the most general. Of all the woods most in use, oak is the chief, being used for frames, stringers, keels, stem and sternposts, deadwoods, deck beams and outside planking. Oak varies very much according to the locality in which it is grown, the soil, and the climate. The best oak is produced in the dryer parts of England, such as Kent, Sussex, Yorkshire, and the midlands. The climate of Ireland and the west of Scotland is unfavourable to the production of the best class oak. Oak should be felled in the late autumn, when the sap is down. Spring or summer-felled oak is worthless, and may always be known by the clean absence of bark from the log. It requires long seasoning, especially when used for planking. Oak lasts practically for ever when kept absolutely dry, or absolutely wet. Alternate wet and dry soon ruin the best of oak. It should therefore be used with care as a material for planking, especially if the yacht is to be hauled out every winter, as even the best of well-seasoned oak will "come and go" very extensively under these circumstances. In carvel work of 1 in. or less thickness the planks should be narrow and thoroughly well protected by paint or varnish. Oak weighs, when dry, about 50 lbs. per cubic foot.

Elm is used for keels, transoms, and, in large vessels, for the lower planking. There is no wood that is more difficult to preserve when used as planks for small craft, unless it is thoroughly well seasoned and constantly protected. Large elm logs, which are used for keels, are liable to turn out full of bad places, which are only discovered when the log is sawn.

Wych Elm, or the large leaf elm, is of smaller growth, and is a more graceful tree, with less rugged bark. It is, when carefully selected, good for bottom planking of small boats, and is of a tougher nature than the common elm. Unfortunately it is not very available, growing in but few places in any abundance, except in some parts of Yorkshire. I have seen some boats built of wych elm which were sound after twenty years' use, but such cases are rare. One quality not generally known is the rapid deterioration of elm when transferred from salt to fresh water, or *vice-versa.* Many clinker-built beach boats in the South of England are planked with elm, which stands the rough work of hauling up and down a beach very much better than oak. Elm should be kept under water when in stock.

Larch, like oak, depends for its quality upon soil and climate. When grown in wet and loamy districts its rate of growth is rapid, but the wood is soft and uneven in quality, and very coarse in texture. The wetter the climate the coarser the growth. The best larch has only a small portion of sap, the heart wood being of a beautiful reddish colour, of smooth texture and without very many knots. If of first class quality it is often used for planking. Larch is tough and flexible, and is therefore useful for shelves, stringers and deck beams. Unless well over 1 in. thickness it is not quite so good as other woods for planking when used for carvel build. For clench build it is as durable as oak, though not so beautiful in appearance, and far better than elm for withstanding decay. Ordinarily trees from 9 to 13 ins. diameter give the best timber. The greater proportion of Scotch fishing boats on the East Coast are larch built on oak frames. Seasoned larch weighs about 37 lbs. per cubic foot.

Fir is not a good wood, whether English or Baltic. For many years past it has been impossible to obtain red or yellow deal. (which is fir) of a sufficiently good quality and free from sap fit to be employed in yacht building. Larch should always be specified in preference in plain jobs.

Teak is *par excellence* the premier wood in yacht work. Its qualities of durability—indifference to wet and dry changes—and its strength, indicate its fitness for planking, and also for companions, skylights, covering boards and king planks. Against its virtues there are only two faults—its weight and a certain shortness of grain. The best qualities of Rangoon teak are now very costly, and this fact has caused many inferior grades of so-called teak to be placed on the market—such as " Yang " and " Eng " teak—which are hardly teaks at all. Some African teak[5] has qualities and appearance similar to good class Burmah, but little is actually known of its lasting properties, and it is less true in grain, and consequently harder to work. Teak is said to be impervious to the attacks of worm (*teredo navalis*), but recent experience has disproved this statement. Teak stands variations of temperature extremely well, and is therefore excellent for topside plank, and when used for decks on large craft may be of less scantling than any other wood. Teak weighs about 55 lbs. per cubic foot when of the best quality; the inferior kinds are much lighter. Recent quotations for the best quality Rangoon teak reach 18s. per cubic foot.

American or Canadian Rock Elm is a valuable wood for yacht building purposes. If of good quality it is excellent for keels, bottom planking, shelves and stringers. Nearly all steamed timbers are of American elm, though both English ash and acacia are used for this purpose when obtainable. Acacia is said to be better than any elm, and is largely used in France, where the tree is fairly abundant. Many small boats are timbered with ash, which, when well protected, is durable and tough. American elm is a heavy wood, stringy and tough in texture, working well and taking good finish. Like English elm, it is, in large logs, liable to serious concealed defects, which can only be discovered when sawn, hence it is necessary to watch carefully its use for keels, as unscrupulous builders will cut out the defective parts and insert what are called " graving pieces " to conceal the mischief. Wood with blackish streaks and spots should be rejected. Unless well pro-

tected by varnish or paint American elm soon deteriorates. American elm weighs, when dry, from 35 to 38 lbs. per cubic foot.

White oak is an American wood very largely used in yacht construction in that country. If of good quality it is an excellent wood for steamed frames, deck beams and carlines. It is of a straighter grain than English oak, somewhat lighter, rather more elastic, and better to work. It is not difficult to obtain, but requires long seasoning, like its English namesake, before use. It should not be used for planking, as, between wind and water, it rapidly perishes.

Pitch Pine.—The best pitch pine, or " hard pine," as it is called in America, comes from the neighbourhood of Pensacola. It is an excellent wood for planking, shelves, stringers and deck beams in all but first-class jobs. In the best qualities it is straight-grained and free from knots, and makes fine planking if of a fairly " mild " nature, *i.e.*, free from excess of tar and swirls. It is a heavy wood (a dry sample will sometimes weigh over 43 lbs. per cubic foot), but it does not shrink greatly when exposed to variations of wet and dry, being in this respect superior to oak. A pine very similar in appearance, but far less durable, called " North Carolina," is to be met with occasionally. It is not to be compared with genuine pitch pine for durability and standing qualities. Good pitch pine has increased in price very much of late, and much inferior stuff has in consequence been put on the market.

Kauri Pine.—A New Zealand wood of a very valuable character and of great use in yacht building for deck planking, skin planking and cabin fittings. It is somewhat heavy if of good quality, from 33 to 35 lbs. per cubic foot, and takes a high finish. It has almost superseded yellow pine for deck planking, and, although its weight compares unfavourably with this wood, its wearing qualities are much greater. It is said to shrink longitudinally (an almost unbelievable assertion), and to be subject to rot when used for outside planking of light scantling. After a close experience of about fifteen years, I have not found this to be the case with well-seasoned stuff. Its chief defects when used for deck work are its slippery surface, especially when oiled

5 Afrormosia

or varnished, and the curious dark brown stain it exudes, unless varnished, when subject to heavy and continuous rain. This stain, unless immediately removed from the topsides, becomes very hard and penetrating, making an almost permanent blemish on white enamel. It is an excellent material for topside planking, taking paint or enamel well. It is essential that this wood be well seasoned before use.

Yellow Pine.—Almost unobtainable in the best qualities, that which now comes into the market being used chiefly for pattern making. It is called " white pine " in America—its native home. When it was plentiful, some thirty years ago, it was used for the planking of yachts and fishing boats, and for the deck planking of all yachts. Though a soft wood, it is remarkably durable when well protected by paint or varnish, and many Loch Fyne boats planked with this wood are still in existence.

Mahogany.—The best Honduras mahogany ranks next to teak for durability, and has the advantage of being considerably lighter. Consequently it is much used in high class work for topside planking, deck fittings and cabin panelling. But it is extremely expensive in the best qualities, and many different and inferior kinds are in use. Mahogany weighing less than 35 lbs. per cubic foot should not be used for planking or deck fittings, as the lighter kinds of mahogany are very absorbent of water, and are little better than cedar in strength; good pitch pine or Kauri pine being preferable. " African " mahogany[6]is not a true mahogany, and is only fit for cabin fittings. It looks well when stained and polished, and is

6 Khaya

good enough for this purpose, being light and easily worked. Skylights and companions on racing yachts are nowadays almost invariably of hard mahogany, and its use for these purposes is increasing in cruising yachts. But for strength, durability and resistance to damage teak is to be preferred for all deck work.

Oregon Pine is chiefly used for spars, being obtainable in long lengths clear of knots, and is of a tough and elastic nature. The best kinds are of dark yellow or reddish colour. It is occasionally used for deck planking, and for topsides, but for either of these purposes it is inferior to the best Kauri. It varies greatly in quality; some kinds, especially the red, being coarse in texture, and having a great proportion of sapwood. Logs of 60 and 70 feet length are not uncommon. Its weight is from 30 to 35 lbs. per cubic foot.

Cedar was the early name for mahogany on its introduction to Europe in the sixteenth century. Nowadays it is applied to inferior qualities of mahogany, in which are included the terms " Baywood " and " Juniper," or pencil cedar. All these woods are occasionally used in the construction of yachts, but chiefly in the smaller classes. It is a light, soft wood, easily bruised, and very absorbent, unless the outer surfaces are well protected by varnish. It is, in common with African mahogany, now much used for cabin fittings, where its lightness and pleasant colour have advantages over other woods. Its weight varies from 28 to 30 lbs. per cubic foot, the lighter kinds being very unsuitable for the planking of cruising craft.

SHEILA II – *an anchorage at Lunga, Treshnish Isles.*

CHAPTER 3

The Designs

THE DESIGNS

In making the following selection of Albert Strange designs, we have chosen to avoid duplication with the designs published in John Leather's book, *Albert Strange, Yacht Designer and Artist*. The selection is nevertheless representative of the variety of Strange's output and spans the years from 1896 until his death in 1917. It should also be stressed that, with this and the earlier selection, we have now reproduced almost all the *Yachting Monthly* designs together with a fair number found in books long out of print. Other designs are found in the collection at Mystic Seaport Museum.

A number of the designs were accompanied with Strange's detailed specifications and scantlings when originally published. In a few cases, tables of offsets also exist. A full list of references for each design can be found in the Bibliography, page 263f., to enable further study.

The design numbers where available, are Strange's own. the majority of the comissioned designs (irrespective of whether they were built) were numbered. The competition entries and design exercises, as well as some unfinished designs and sketches, generally lack numbers. (It will be noticed that *Cloud* only received a number when the design was modified for an actual client, four years after it was originally completed for a competition.) The design numbers, taken in conjunction with their dates, provide some indication of Strange's output. For example, Nos. 96 to 108, together with the un-numbered *Desire*, all fall within the year 1909, indicating that Strange was producing designs at the rate of about one a month during this period, as indeed he did for a considerable number of years – an impressive record considering his other commitments.

The designs are arranged here in broadly chronological order, although it is hoped that the occasional exception will allow easier comparison or contrast to be made between particular boats.

Where necessary, the dimensions appearing under the name of each design have been converted from feet and decimal fractions of feet, to feet and inches. Elsewhere all dimensions appear as originally printed. The displacement and Thames Measurement tonnage (TM) is given only where originally included with the designs.

(For a metric conversion table see Appendix 4, page 275.)

LOA, or Length Overall, in all cases refers to the overall length of the hull and does not include bowsprits, bumpkins, overhanging booms, rudders or any other appendage to the hull. Until the last decade or two, this practice was universal. It would not be necessary to make this point were it not for the quite senseless corruption of the term, perpetuated by those who believe they use it accurately by taking it 'literally'. Being neither universal nor defined, this modern practice has sadly and quite needlessly created confusion over a convention which has been used to describe many thousands of vessels for the best part of a century throughout the western world, and which is still used in the IOR Rule and by those who appreciate the value and convenience of upholding this usage.

Geo. Holmes   *YM Sept 1916*

MONA 1896

LOA: 23ft 0in	LWL: 17ft 9in	Beam: 6ft 9in	Draught: 1ft 7in/5ft 0in
		Ballast Keel: 8cwt	TM: 3 tons

The drawings of *Mona* here reproduced were published in the 1904 edition of the Humber Yawl Club *Yearbook*, together with a commentary put together by the editor from notes provided by her owner, Mr. Webster. *The Yearbook* relates that

his attention was first seriously directed to small yacht sailing and cruising, by his coming across the canoe-yawl Cherub *on her voyage to London in '93, when she passed through Lincoln. A hasty sketch of a craft suitable for inland sailing, made on the back of a telegraph form by the Skipper, was carefully preserved and studied and, in the course of two or three years, bore fruit in the determination to build. The resulting craft was constructed by her owner from drawings provided by Mr. Strange and, although subsequently considerably altered both in the upper works and by the substitution of a lead for an iron keel, the main form remains substantially as at first designed.*

Cherub's voyage to London was recounted by Strange in *The Yachting Monthly* under the title 'A Single-Handed Cruise in the North Sea' (YM, February 1908, also JL p.37ff.) Strange, too, recalls the Lincoln encounter:

… we were visited by a good many people who were interested to see a "sea-going" sort of boat so far from her proper element. Cherub *had to be explained and I made several converts – one very notable one, who is now one of the hardest sailors on Trent and the district.*

Webster evidently liked to race, and did so with some success in *Mona*. He also cruised her much further afield than originally planned; on one occasion making a substantial coastal passage as far as Bridlington. Such voyaging would account for the many modifications he made to the boat.

One of these was to the sail plan, as indicated in the drawing, which Webster changed from gaff mainsail, topsail and jib set flying, to what he refers to as a 'bafter'[1] mainsail and jib of considerably greater area. Both sails of the new rig were fitted with roller reefing.

Webster's alterations to 'the upper works', are explained in an entertaining way by a later owner of *Mona*, who was evidently somewhat mystified by them. The following is quoted from an article he wrote under the initials D C M, in *The Yachting Monthly*, December 1947, in which he remembered his early sailing adventures in the late 1920s.

She was a grand little boat, but lacked freeboard. She could have done with another 6ins. Mona *was always something of a mystery, for she had already been built up 4ins to 5ins more in the bows. Yet her lovely decks showed no sign of having been tampered with, but right aft there were spacers between the original beams and the deck and all along her timbers had been lengthened by short pieces laid alongside the main parts.*

The only explanation I ever heard was that her designer was under the impression that she was to be used on inland waters, whereas her owner wanted her for the Humber Estuary. Whatever the reason, she still had less than a foot of freeboard and she was fast; and so she was also wet. Lord, how wet she was! But a grand boat just the same. I spent several holidays in her, pottering about the Wash, often in places where no boat should be, but my ambitions were growing and I wanted something bigger. I wanted a robust boat, on which hobnailed boots could be worn, for some of my crew appeared to favour this kind of footwear and I wanted a greater degree of comfort. At the end of a day of cold and wet, I found that heaving up a lifting cabin-top and getting to work with a mop, before I could even think of a meal, was discouraging.

I found, too, that playing a sort of musical chairs when more than one wished to enter the cabin was inconvenient.

Mona's centreplate was in the way; it was the old-fashioned triangular type. In a seaway it made disgusting noises, as of an elephant in distress and gushed water all over the cabin.

I wanted a man of a boat. Mona *had an ability to get to windward in conditions that should have overwhelmed her, but in doing so she qualified for the submarine service. After a day of wind and water that the Editor described as 'too distressing for our readers', I was decided. I bought a smack.*

Mona is Strange's seventh yacht design (excluding designs for model yachts), of which we have any tangible evidence and probably the earliest still in existence. The design is an early example of his versatility. The previous known designs are early canoe yawls, yet here he demonstrates his ability to design an entirely different type of vessel.

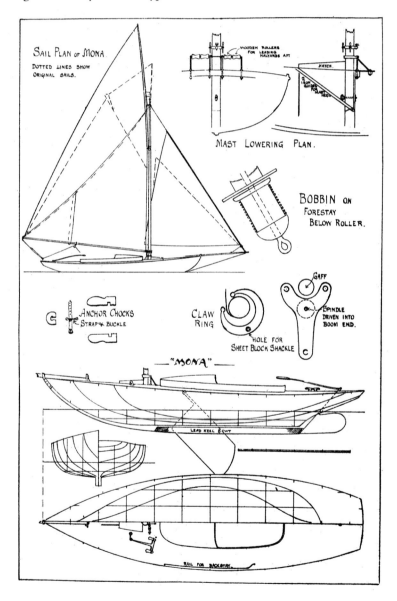

1 The term bafter distinguished what we now call a gunter, or gunter lug, from the sliding gunter. With the latter, the yard was hoisted parallel to the mast on two sliding bands, whereas the bafter yard had single jaws pivoting on the mast to which it was held almost parallel by the peak halyard.

TAVIE II 1896

LOA: 22ft 0in	LWL: 19ft 0in	Beam: 6ft 4in	Draught: 2ft 0in/4ft 3in
Sail Area: 260sq.ft		Ballast Keel: 11cwt	TM: 3 tons
		Centreplate: 235lbs	

This canoe yawl was designed by Strange in February 1896 for Dr. John Hayward of the Mersey Canoe Club. She is very similar in concept to Strange's own boat, *Cherub II*, of 1893, though rather larger and fitted with the 'lever' type of centreplate referred to on page 122. Although proof is still lacking, it is very likely that these two men first met when Strange was teaching art in Liverpool around 1880.

Tavie II was built by Bond in Birkenhead, Merseyside, in the same year that the design was completed, and her owner described her in a letter to *The Field* magazine, of which the following is an extract.

> Tavie II *is a single-handed cruiser of the canoe-yawl – or canoe-yacht – type, ... excellently built and finished by Sam Bond of Birkenhead.*
>
> *The boat draws a little over 4ft. with the centreboard down. There is about 3cwt. of inside trimming ballast in lead pigs. The cabin is quite comfortable for two to sleep in and, with the hatch raised on the side shipping pieces, the accommodation is even luxurious. For dryness and comfort it beats any tent I have seen, and requires no laborious setting, reefing, or furling by a tired man over a wet cockpit. The boat takes the ground well, although, with her rather sharp floor, this is somewhat more pleasant if she is on her legs ...The cockpit is bulkheaded from the cabin and from the large, roomy after locker. When the cabin is shut up, therefore, the boat is a lifeboat, as well as being practically uncapsizable.*
>
> *... It only remains for me to add that, after three months' experience of the boat, I believe I at last possess the ideal single-handed cruiser. All halyards and sheets lead to the cockpit, as does the fall of the centreboard tackle and they are readily worked therefrom ... I have double tiller lines rove forward, but have not required to use them. Though not built for racing and not canvassed to her capacity, the ship is no slut even in light airs; in a breeze she is remarkably stiff, dry, and weatherly. She will sail under mainsail alone, and on two heavy squally days with considerable sea, she was found to work successfully with care under jib and mizen. I consider her chief virtue to be comfortable running before waves – a virtue which was once of great advantage running over Conway bar ... In a fresh breeze she prefers the mizen stowed.*
>
> *With a crew of two the boat would stand a lot of weather. I have been in her from the Mersey to Conway, Moelfre Point, Beaumaris, Caernarvon and back. [These are all coastal passages of 50 miles or more in exposed waters.]*
>
> *The only faults I have yet discovered are that she carries too strong weather helm, and that she can [only] with difficulty be got to balance and steer herself while the skipper cooks or monkeys. The first fault can be easily remedied by one or more of the well known methods; the second, a drawback in a single-hander, is somewhat inherent in the design of a short boat with a cutaway forefoot.*
>
> *For nearly 20 years I have been a single-hander and during that time, I have owned and sailed various craft from Rob Roy canoes to a four-tonner. As a result of this experience, I can recommend a similar boat to my present one to anyone with similar tastes.*

The most well known method of attempting to cure excessive weather helm is to increase the 'lead' between the centre of effort of the sail plan and the centre of lateral resistance of the hull. Studying the lines drawing, it appears that the designed difference is very small. It would be interesting to learn the effect of removing a couple of cloths from the mainsail and increasing the area of the jib.

However, it appears this experiment was never tried. She was shipped out to Abadan before the Second World War and someone who sailed her there in the 1950s again remarked on the weather helm.

Before leaving *Tavie II* and the relative merits of lifting cabins over boom tents, to say nothing of Dr. Hayward's genuine enthusiasm for his ideal cruiser, mention must be made of the yacht he commissioned 13 years later in 1909. This was *Vanderdecken*, an 87ft. auxiliary yawl of 80 tons, designed and built for him by Stow & Son of Shoreham, Sussex. In the early 1990s this yacht was tragically lost while on a transatlantic passage to the U.K., where she was to have undergone a major refit.

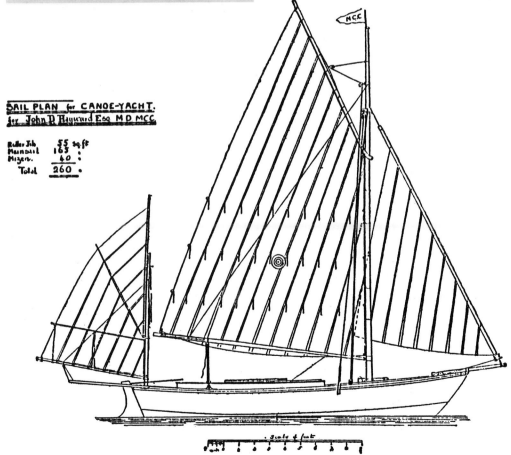

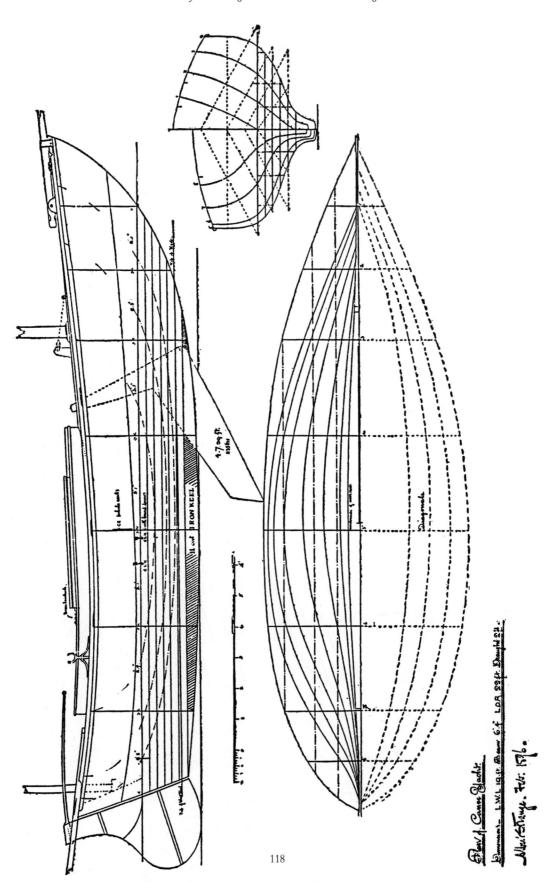

WENDA 1899			No.45
LOA: 24ft 9in	LWL: 19ft 3in	Beam: 6ft 5in	Draught: 2ft 3in/5ft 0in
Sail Area: 295sq.ft	Disp. 1.5 tons	Ballast Keel: 12.6cwt	Centreplate: 180lbs

Wenda was designed for Edmund Bennet, an architect who lived in Gravesend on the lower Thames, where Strange was himself born and spent his youth, although he had been living in Scarborough for over 15 years when he received this commission. *Wenda* is designated a 'Fast Cruiser Canoe Yacht' on the drawing. Her plans were published in the fifth edition of H C Folkard's book *The Sailing Boat*, in 1901. Folkard describes the 'leading requirements' of the design...

> *the yacht should be constructed with as light a displacement as possible consistent with immunity from capsizing. With fair accommodation for a crew of two persons, but as regards management, to be within the power of one to work a passage single-handed, or to enjoy an afternoon's sail. Speed was to be kept in view, but at the same time the extent of the sail area was to be very moderate. In fact the instructions were that the boat should, as far as consistent with other special requirements, be 'a good all-round cruiser, capable of making coast passages and yet be a good performer on the river'. The necessity of light displacement was imperative, as the boat was to be capable of being shipped to foreign ports as inexpensively as was compatible with the length.*

Wenda is a most beguiling design. Her sweet lines and generous overhangs make her appear bigger than she is. In fact, her waterline length, beam, draught, and least freeboard are almost exactly the same as those of *Tavie II;* and below there is only sitting headroom for someone under about 5ft tall. However, with her extra overall length and light displacement, she would be a versatile craft in the short steep seas and tides of the Thames estuary. Her low-aspect sail plan is well matched to the light and slender hull, and the lug mainsail (much favoured for racing boats at this time) is quickly and easily reefed.

It is a sad fact that there is no record of *Wenda* having been built. However, when her lines were published in both *Cruising World* and *WoodenBoat* magazines in the USA in the 1980s, they generated such interest that *WoodenBoat* commissioned Phil Bolger to copy the lines on a larger scale, draw up a table of offsets, and make construction drawings. Bolger wrote that he did so acting strictly as Strange's draftsman and that *Wenda's* hull form and deck plan were as close to the surviving tracing as he could make them.

A number of boats have now been built to these plans.

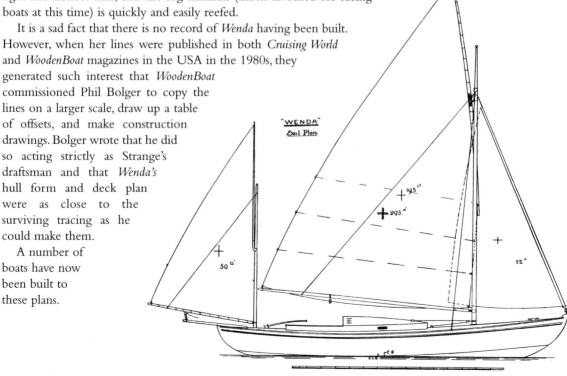

"WENDA"
Sail Plan

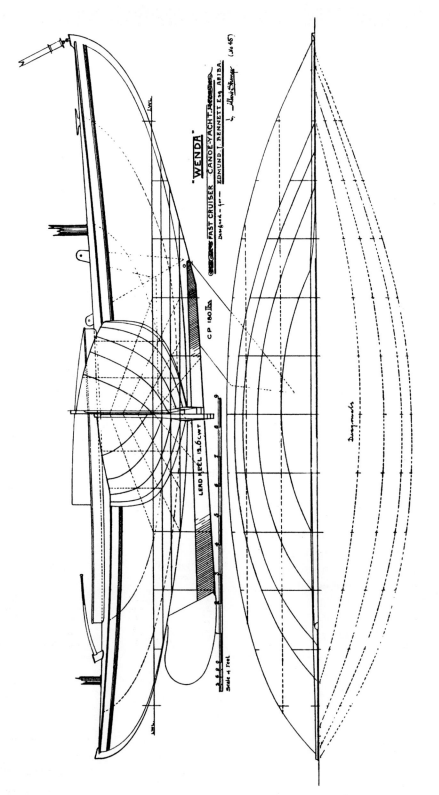

WENDA *canoe yacht.*

DUNLIN (Solway Class Boat) 1908   No.87

LOA: 21ft 6in	LWL: 19ft 6in	Beam: 7ft 0in	Draught: 2ft 3in/4ft 8in
Sail Area: 298sq.ft	Disp. 1.98 tons	Ballast Keel: 11.3cwt	

THE SOLWAY CLASS BOAT

Sir, – A correspondent in your last issue asks if there are any survivors of the 19ft square-sterned boats designed for the senior Clyde Club by the late G L Watson. I fear that very few exist today, though one or two may occasionally be seen still serving as good cruisers at out-of-the-way ports in the West Highlands. The class was a thoroughly good one of its type, now, alas, rapidly being forgotten in the general desire to possess a boat that must at any rate look fast and turn on her heel with abnormal celerity – a qualification not always required in a cruiser, in whom a maidenly sedateness of movement occasionally brings comfort to the owner.

If you should care to publish the accompanying lines of the Solway Class boats they may perhaps be of some assistance to your correspondent and perhaps others who have to keep their boats in places where the short-keeled type of deep draught cannot comfortably exist.

The conditions given for my guidance in preparing the design called for a boat of light draught and ability to take the ground every tide. The boats were required to be stiff and dry, capable of being used as weekend cruisers and for fishing and shooting expeditions. The rig was to be small and fit for single-handed work, and they were not to be expensive in build or fittings. The very strong tides in the Solway favour a light-draught boat, as almost anything will get to windward on a fair tide there, and hardly anything can beat to windward over a foul tide. The plate is therefore very small and unobtrusive. In other places where tides are easy, a deeper and larger plate would be better, but on the Solway the amount given will be found ample, as it would also on the Humber and Severn. With their long keels the boats should run and reach easily and steadily in the broken tidal water of the districts. The boats are clench built of larch with bent timbers and grown floors of oak. Decks are canvas-covered, as is also the hatch. The class will be built this winter, and I have no doubt that the secretary, Mr Wilson H Armistead, of The Scaur, Dalbeattie (by whose kind permission I am allowed to send you the lines), would give anyone intending to build particulars as to the builders and probable prices.

Albert Strange

November 1908 was a good month for Albert Strange. As well as having his winning design *Cloud* illustrated in *The Yachting Monthly*, he also had the above letter published, together with the lines and sail plan of the Solway Class Boat.

In his letter, Strange writes of the planned building of the class as a certainty but, as far as is known, only one example, *Dunlin*, was in fact built, for Wilson H Armistead, who used her for cruising about the Solway Firth in southwest Scotland. Armistead was a keen wildfowler and also indulged in some amateur trawling. For this latter purpose he lengthened the bowsprit and increased the total sail area to 350sq. ft.

Although Strange is best remembered for his development of the canoe stern, he acknowledged that a transom stern provides the maximum amount of room in the cockpit of a small boat, and produced a number of designs of similar size to his Solway Class. He also sketched a yawl version of this design, with a fixed cabin, for another prospective client.

It was reading Strange's account of his cruises in *Cherub II* (JL pages 19f., 37f., 85f.) that inspired Armistead to build 'a somewhat similar boat'. He was pleased with *Dunlin*, but an awkward accident a couple of years later prompted him to write to *The Yachting Monthly*,

… I have just been having trouble with the plate of my 4-ton sloop, Dunlin*…. For a long time I escaped trouble, till I began to think that so far as* Dunlin *was concerned the plate was all that could be desired. I had been frequently aground, but always got clear without inconvenience.… When the plate*

is down it will give warning of shoal water . . . Several times I forgot to lift the plate when coming ashore, and still I escaped trouble, though I daresay I deserved it . . . What subsequently happened makes me distrust centreboards forever, unless an improvement can be made in their mechanism. I was beating up a narrow channel with steep sides, with the plate down and a fairly strong tide against me. Standing too far across, the plate touched. Immediately Dunlin's bow was jammed against the bank by the tide, and the centreplate so firmly fixed in the case that it was impossible to move it, even with the help of the throat halyard. As the tide ebbed, the yacht lay over towards the bank and the pressure on the plate increased till at last, when she was dry, there must have been about two tons on it and of course it bent badly.

Strange, with long experience of the centreplate – not to mention that he was the designer of the boat – was swift to reply, couching his criticisms of what he considered to be Armistead's very bad habits in the most elegant terms,

Sir, – Mr W H Armistead surely has had a longer period of immunity from trouble with his centreplate than his conduct entitled him to. The list of sins of omission and commission he gives is amply long enough to account for any severity on the part of outraged Chance.

Nevertheless I sympathise with him – a bent plate is a great nuisance. Like a drunkard it offers little security against a relapse . . . I hope he will soon have the plate working again, and that he will stick up in some part of his cabin the following motto: 'Eternal Vigilance is the price of Freedom'.

The plate in Dunlin is the most unobtrusive form possible ... But if anything happens to it, it is perhaps the most difficult to deal with because of its length – a very trifling bend makes it unworkable.

The best plate of all is the lever plate [see below], taking up little valuable room, but certainly the easiest to work and the least liable to accidents. It should, however, never be less than ⅜in thickness and it is rather heavy in comparison with its effective area. It was first used in Tavie II in 1896 [see page 118] and then Eel was built with it in 1897. It has rarely given trouble in either of these boats. But then the owners are careful men ...

Armistead felt he had been misunderstood:

I accept with humility Mr Strange's rebuke concerning my sins ... but I would point out that the disaster had nothing to do with any of these. I really doubt whether the excellent motto suggested would help me, for I have observed with sorrow that the wisdom of all the ages has to be acquired by each of us through experience and I have never yet met the man capable of 'Eternal Vigilance'.

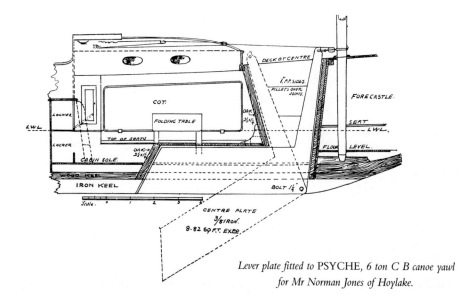

*Lever plate fitted to PSYCHE, 6 ton C B canoe yawl
for Mr Norman Jones of Hoylake.*

Armistead's question, whether a system of rollers inside the case could not be devised to enable the plate to be lifted even while it was being subjected to a lateral load, was never properly answered. He also asked why wooden boards were not more often used, as in the USA. Strange replied that they took up too much space, without explaining that this was due not only to their greater thickness, but also to their shape, which the nature of the material would not allow to be of either the Solway type or the lever type. Modern epoxy composites make the board a more attractive proposition. A further discussion of centreplates is to be found on page 58f.

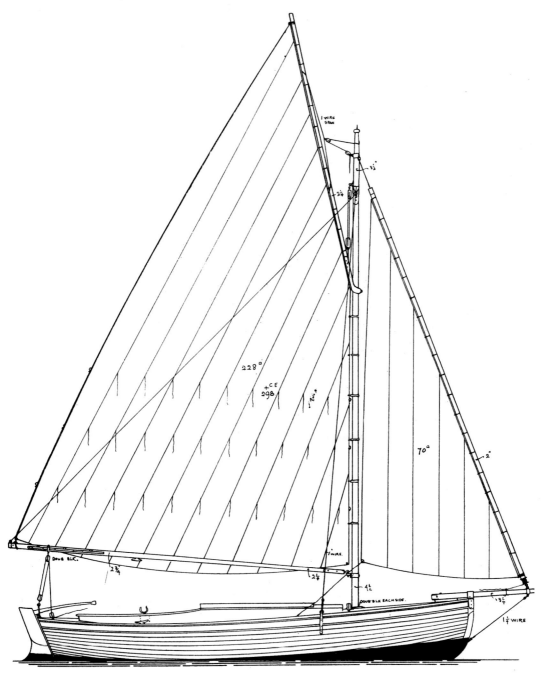

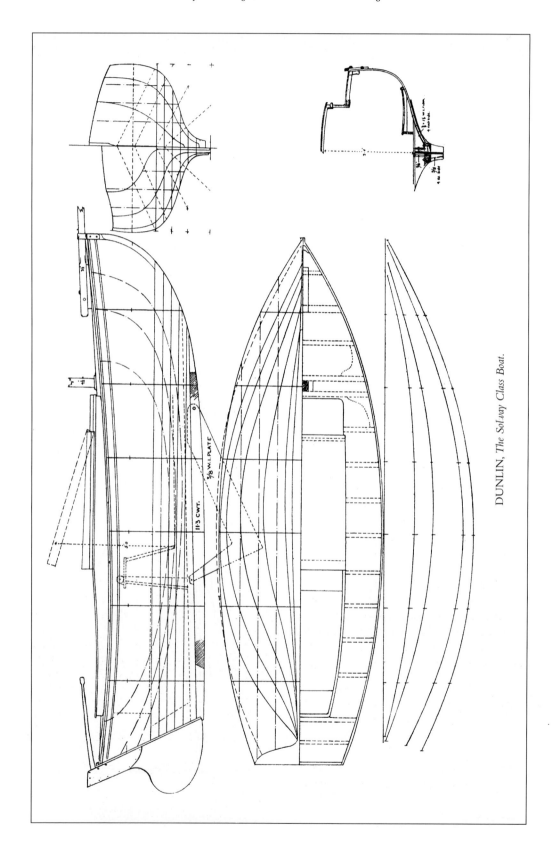

DUNLIN, *The Solway Class Boat.*

QUEST II 1907

LOA: 44ft 0in (approx.)	LWL: 36ft 0in	Beam: 11ft 7in	
Sail Area: 1,040sq.ft			TM: 21 tons

This handsome ketch is only known to us through the accompanying photographs, sent to *The Yachting Monthly* by her owner C W Adderton, with the following brief letter,

Sir, – In view of the controversy about canoe sterns and ketch rigs, together with the great interest taken in your cruiser designs, would you care to publish the enclosed photographs of Quest II?

Having gone through the various stages of ownership of converted fishing boat, 10-ton cutter and 12-ton yawl, I claim now to have reached the 'ideal cruiser' type.

The editor appended a note to this letter, 'It may interest our readers to know that this boat is the prototype of *Cloud* and, like our winning design, was designed by Albert Strange.' (*Cloud* appears on page 127f.)

C W Adderton had been a pupil of Strange's at Scarborough School of Art. The 12-ton yawl to which he refers, was the first *Quest* (built in 1903 on the Isle of Man, designed by a Mr Hamilton), in which Strange cruised with Adderton in the summer of 1906. He described her as a 'sweetly pretty yawl' and had a great regard for the yacht and the efficacy of her yawl rig. He described the cruise in an article in *The Yachting Monthly* some years later (see JL page 105f.).

In about 1904, Adderton had taken the first *Quest* on a cruise to the Solway Firth. It was here that he met Wilson Armistead, for whom Strange later designed *Dunlin*, (see page 121f.).

In 1911, Armistead wrote a long article in *The Yachting Monthly*, describing the Solway Firth from the cruising yachtsman's point of view. It was illustrated with many fine paintings by Adderton, whose arrival with *Quest* in Armistead's home waters had created something of a sensation, at a time when 'yachting' was virtually unknown in the area.

Never before had a smart, modern, 12-ton yawl visited the little village and, naturally, she created considerable interest .... Not only did her model seem unusual, but the clean, well-varnished spars, spotless sails, teak rail and deck fittings, with brightly polished brasswork, were things which called for remark, where tar and paint had hitherto satisfied the aspirations of owners of local craft.

All this happened six years ago and though much progress has been made in the nautical achievements of the Water of Urr, the coming of Quest *will always be remembered as an epoch-making event.*

Quest II was about 44ft overall, 36ft LWL and 11.6ft beam. She was built by A M Dickie at Tarbert. In 1913, Adderton went back to Dickie for *Quest III,* a rather larger yacht of 39ft LWL, this time built to a design of Peter Dickie, who was to produce a number of fine canoe-stern cruising yachts in subsequent years from the yard he later owned at Bangor, North Wales.

In the obituary of Strange carried by *The Yachting Monthly* in 1917, 'His old friend and former pupil, Mr C W Adderton, writes of him, "No man ever had such a dear friend and I shall miss his constant sympathy and help more than I can express". '

QUEST II, *at moorings.*

QUEST II, '*all told.*'

CLOUD 1908 & 1912				No.125
YAWL	LOA: 38ft 9in	LWL: 30ft 0in	Beam: 9ft 7in	Draught: 5ft 3in
	Sail Area: 846sq.ft.	Disp: 11 tons	Ballast keel: 4 tons	TM: 13 tons
			Internal: 1.15tons	
KETCH	LOA: 38ft 9in	LWL: 30ft 0in	Beam: 9ft 7in	Draught: 5ft 0in
	Sail Area: 777sq.ft	Disp: 10.13 tons		TM: 12.4 tons

When this design won the 'Sixth *Yachting Monthly* Designing Competition' in November 1908, prize £10, she was drawn with a ketch rig, incorporating many labour-saving ideas for single-handed sailing. These included a fore horse for the staysail sheet, roller-furling gear for the jib, roller reefing main and mizzen, main topping lift and peak halyard led aft, and tiller lines on snap hooks led forward to the foredeck.

The competition called for:

A single-handed cruiser which shall not exceed 30ft. on the waterline. It is generally recognised by single-handers that a boat of considerable size is often easier to handle and manoeuvre than a small one … Needless to say, the boat should be wholesome in type, the accommodation good and rig simple. The working of the gear should be clearly defined and details supplied, together with specification.

The deadline for entries was the first post of 5th October. Since the above announcement was printed in the September issue, it is remarkable that a total of 22 sets of plans were submitted in the short time available. Unfortunately, no complete list of competitors' names was published, but among designs commended were examples by F C Morgan Giles, then at the start of a long career as a naval architect, and J Pain Clark, a well-known amateur designer.

Albert Strange based his winning entry, *Cloud,* on a 44ft-LOA canoe-sterned ketch named *Quest II,* which he had designed during the winter of 1906/7 for his friend and former pupil C W Adderton (see previous page).

The judges commented that *Cloud* had excellent proportions and, despite the light draught, had ample stability for her high sail plan. They considered this sail plan to be perhaps a little complicated, but very serviceable and, with understanding, could be simply managed.

Although complex by today's standards, when anything other than a bermudan sloop is regarded as unusual, the judges' comments still make good sense.

Two vessels are known to have been built to this design. The name, *Cloud,* was retained for the first, which was launched from the yard of W E Thomas, Falmouth, in 1912, for Lt A W Gush RN, and G D Stanford.[1] Her draught was increased slightly to 5ft 3in and she was given a yawl rig of 846sq. ft.

The drawings of the yawl version shown here are reduced from originals held at Mystic Seaport Museum, while those of the ketch are from *The Yachting Monthly.* In profile, the yawl version shows slightly more drawn-out ends and displacement increased to 11 tons.

In the autumn of 1932, by which time she was rigged as a bermudan yawl, *Cloud* cruised from Burnham-on-Crouch, Essex, to Mallorca, returning to the UK in the following year. It was a remarkably eventful cruise, the crew comprising, at the start, the inexperienced owner, A J Williams and a highly unreliable paid hand. Williams later carried on single-handed. Ten years afterwards, he recounted his experiences in *Yachting World* and they make hair-raising reading. In one incident, while motor-sailing along the Portuguese coast in a light air, the hand failed to appreciate the

---

1 The name Sanderson appears on some of the drawings of CLOUD, but it is Gush and Stanford who are recorded as her owners in Lloyd's Register.

significance of an approaching thunderstorm and *Cloud* was knocked flat when the squall struck. Her dinghy had been left swung out in the davits and consequently filled to the gunwales, preventing the yacht from recovering, or coming into the wind, and water poured into the cockpit. They eventually managed to get the mainsail down and *Cloud* was able to right herself.

Later, while on passage from Barcelona to Mallorca, *Cloud,* now with only Williams on board, was struck by a succession of violent storms – the 'Levanta' about which Williams had been warned before he set off. The 140-mile passage turned into a five-day fight for survival. *Cloud* lost her jib and then, at the height of the first storm, her sea-anchor cable parted. While lying a-hull, she suffered a violent knock-down which caused the internal ballast to burst up through the cabin sole and it was later discovered, broke six 3 x 3in oak frames and 48 no. ½in bronze floor bolts. The iron tiller was so badly bent that the rudder was inoperable. In trying to bend it back, Williams broke the head of the rudder stock. When the storm abruptly stopped, he managed to work *Cloud* to within a mile of shelter, under mizzen and staysail, steering with an oar. But this 36-hour period ended when a second storm struck as suddenly as the first had ceased. The mizzen boom broke and the sail went to shreds. During this fifth night, *Cloud* drifted helplessly within 100 yards of the Draganora Rock, but was carried around it by a current. In the next sudden lull, Williams decided he must take to the dinghy. It swamped and had to be bailed twice before he could get clear of the yacht. Close to the limit of his physical endurance, he just managed to get ashore at the little harbour of Antrax, where he was able to get a local fishing boat to go out in search of *Cloud*. She was found and towed in just before the next storm struck.

The return trip to the UK was not entirely free of such incidents, but man and boat survived to tell the tale.

Shortly after the Second World War, *Cloud's* next owner sailed her out to Portugal. At the time of writing, she is languishing there in need of a major refit.

Blue Jay, the second boat, was also built by Thomas, but not until 1926. She was built for T N Dinwiddy, a retired architect and experienced seaman. He was a meticulous and thorough man and it is interesting that he should select this 18-year-old design for the yacht in which he planned to cruise extensively through each summer.

With a paid hand as his sole crew, Dinwiddy won the Claymore Challenge Cup of The Royal Cruising Club twice – for cruises to Orkney and a circumnavigation of Ireland.

Blue Jay was rigged as a ketch, but with a larger mainsail and her mizzen mast further aft than Strange's original sail plan. Dinwiddy wrote, 'The ketch rig has proved extremely handy in bad weather, although of course sacrificing something to windward. She has proved fast and well balanced with almost any distribution of full or reduced canvas.'

On Dinwiddy's death in 1945, *Blue Jay* passed into the ownership of the Dents, who owned her for 26 years, latterly living aboard and cruising extensively in the Mediterranean. They also tried a cutter rig and then a yawl. Jean Dent recalls, 'My husband thought *Blue Jay* was finally very much better as a yawl and looked right as well'. In 1999, *Blue Jay* was reported to be undergoing major restoration work in Florida, USA.

CLOUD. The gaff mizzen of the original sail plan has been changed for a bermudan sail.
As a later owner discovered, the davits were not a good place to stow the dinghy in heavy weather (see page 127).
Photograph: courtesy Beken of Cowes.

Beken of Cowes.

"CLOUD~1927."

3128

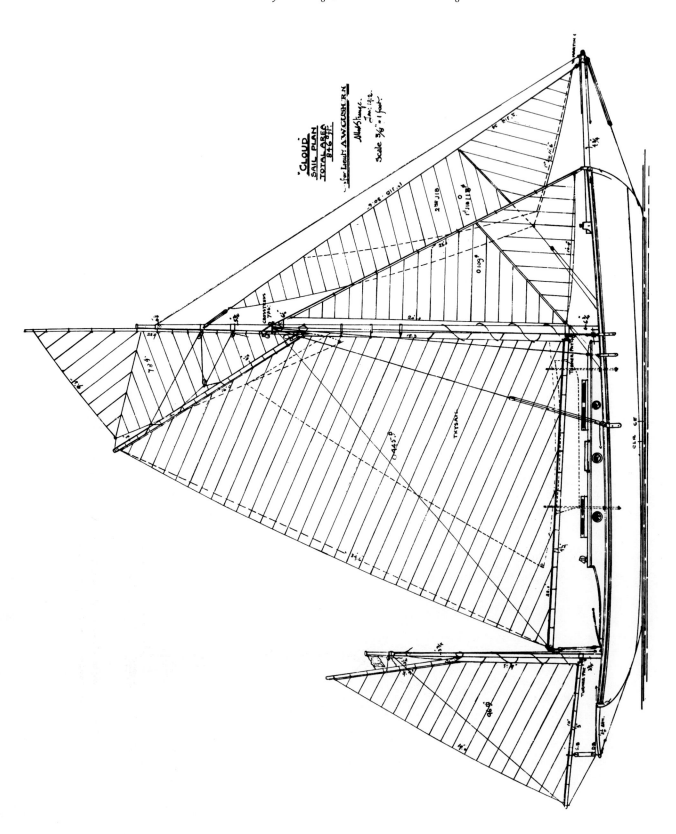

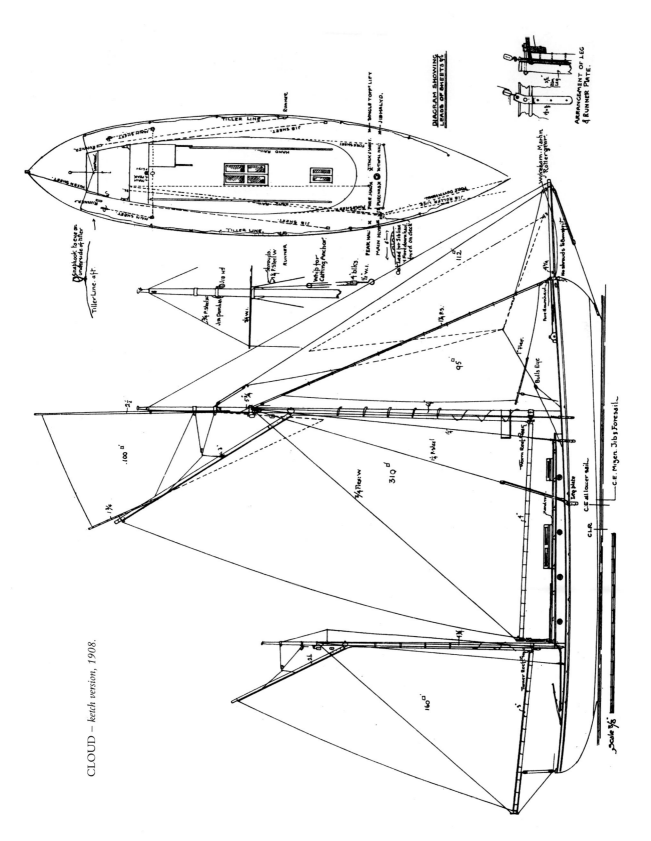

CLOUD – *ketch version, 1908.*

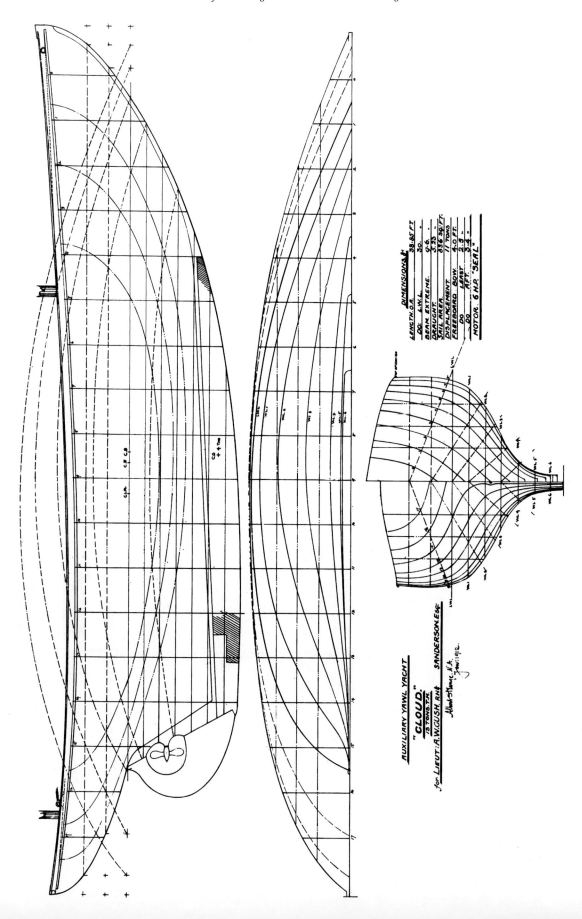

DIMENSIONS	
LENGTH O.A.	38.65 FT
DO L.W.L.	30. "
BEAM EXTREME.	9.6. "
DRAUGHT.	5.25 "
SAIL AREA.	836 SQ.FT.
DISPLACEMENT	11 TONS
FREEBOARD BOW.	4.0 FT.
DO LEAST	2.9 "
DO AFT.	3.4 "
MOTOR. 6 H.P. "SEAL"	

AUXILIARY YAWL YACHT
"CLOUD."
13 TONS T.M.

For LIEUT. A.W.GUSH. R.N. SANDERSON. Esq.

Albert Strange. N.A.
Jan. 1912.

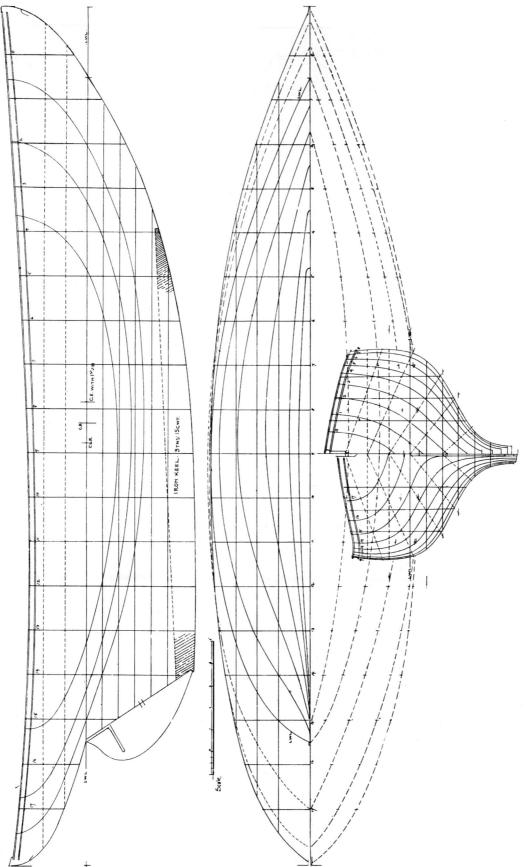

Prize design, CLOUD – ketch version, 1908.

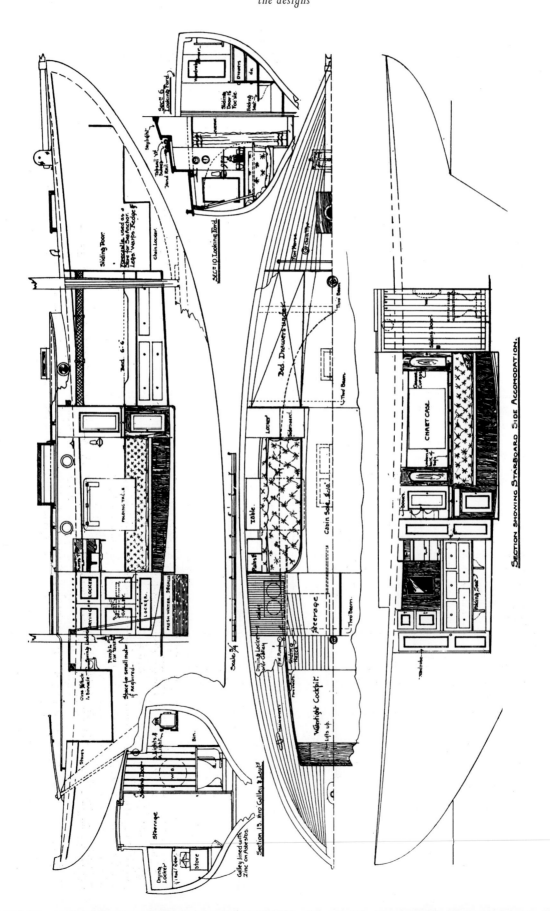

Accommodation plan of CLOUD – ketch version, 1908.

BETTY 1909 — No.96

LOA: 47ft 6in	LWL: 44ft 1in	Beam: 12ft 10in	Draught: 7ft 0in
Sail Area: 1374sq.ft		Ballast Keel: 4.78 tons	TM: 30 tons
		Internal: 8 tons	

At 47ft 6in LOA and 30 tons TM, *Betty* was the largest transom-sterned boat designed by Albert Strange. She was built for Charles Hellyer of Brixham, who had fishing interests in that port, as well as in Hull and was a member of the Humber Yawl Club.

Betty was built by Stow & Son of Shoreham to Lloyd's highest class. The 'midship section drawing shows the hull was to be planked in American elm below the waterline; an unusual choice for planking, (but see page 107).

The commentary which accompanied the publication of the design in *The Yachting Monthly* in 1910, remarks that Hellyer required a yacht in which he could cruise in comfort whilst indulging in deep-sea fishing. This explains the barrel windlass forward of the mast. It continues, 'The transom stern, rather unusual in a yacht of this tonnage, was adopted in deference to the wishes of the owner, in order that she might lie in the crowded harbour of Brixham in the smallest possible space'.

Betty has had a colourful career. When Hellyer commissioned Strange to design the larger *Betty II* of 50ft waterline in 1913 (see JL Fig. 66-69), *Betty* passed into the ownership of Lord Stalbridge, who renamed her *Tally Ho*. The photo on page 140 shows her at this period under racing canvas, with the short pole mast changed to a taller fidded topmast rig.

Stalbridge won the 1927 Fastnet Race in *Tally Ho* and his account of that race in *The Yachting Monthly* is illuminating. The yachts beat down-Channel in a rising gale and at the Lizard, only *Jolie Brise* was ahead of *Tally Ho*. When the former came out of the lee of the Lizard, she was unable to make progress against the WNW full gale. Knowing that by heaving to she would lose the five-hour lead she had established and needed over *Tally Ho*, *Jolie Brise* ran back, speaking to *Tally Ho* as she passed. Stalbridge was not daunted, 'Now was our chance as, knowing from the experiences in a gale in the Bay of Biscay what a wonderful sea-boat the *Tally Ho* was, and also, confident in our sails and gear, we thought that by reefing her down and making things shipshape, we might be able to weather the Lizard'. This they did, with the wind still rising – she was the only yacht to do so that day. *La Goleta* and *Nicanor*, both John Alden-designed schooners and Mr. D'Oyley Carte's smaller, 19-ton cutter *Content*, weathered the Lizard the following day when the wind had eased a little, but *Nicanor* later retired with a broken gaff and *Content* ran in to Cork because of an error attributed to a faulty compass.

Tally Ho just beat *La Goleta* to the Rock, which they rounded in a light air at the centre of the depression, before reaching and running back, in a NE Force 9, which slowly moderated and backed into the NW. In these more favourable conditions, *La Goleta* overhauled *Tally Ho*, but *Tally Ho* was first on corrected time.

At the time, this contest between *Tally Ho* and *La Goleta* was characterised as the hardest fight between two yachts that had ever been sailed in English waters over so long a course and under such heavy weather conditions.

It seems that *Tally Ho* completed more than one trans-Atlantic trip after the Second World War. In 1967, she set out from England heading for New Zealand via the Panama Canal. Her New Zealander owner chartered for a few months in the Caribbean and then sailed on single-handed to Rarotonga, in the Pacific, which he reached in July 1968. Here he was offered a charter to fetch 20 tons of copra from the island of Manuae (Hervey Islands), 120 miles to the northeast. With a young lad as crew, he reached the offing during darkness and hove to waiting for dawn. As they slept, the current carried the yacht down onto the island, where the surf lifted and drove her onto the coral reef and stove in her port side amidships.

She was eventually dragged off the reef, after seven tons of lead ballast had been removed from her bilge and her cabin filled with empty oil drums. As she came off, she rolled over and was dismasted, losing her rudder and bowsprit. But the drums kept her floating just awash and, in that condition, she was towed the 120 miles back to Rarotonga, something of a tribute to the strength of her original deck construction.

In Rarotonga, she was rebuilt over a period of years, during which time she changed hands and eventually found her way, via Hawaii, to the west coast of America. There, with aft wheelhouse and twin trolling poles rigged on the mast, she now earns her living fishing for tuna and salmon out of Brookings, Oregon. She was reported to be for sale in 1995.

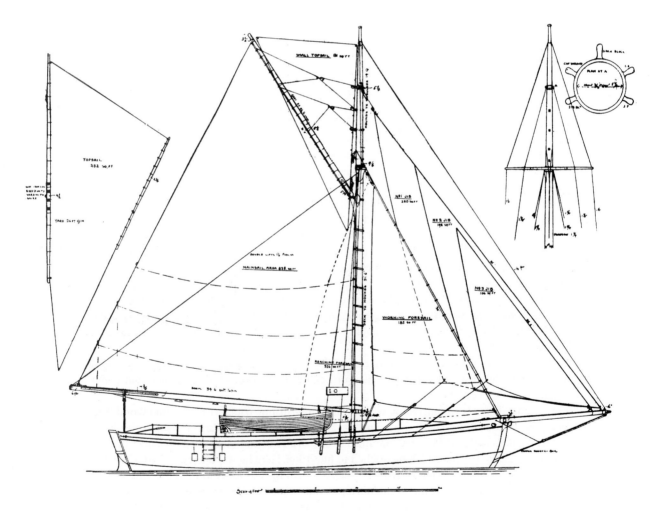

Auxiliary cutter BETTY.

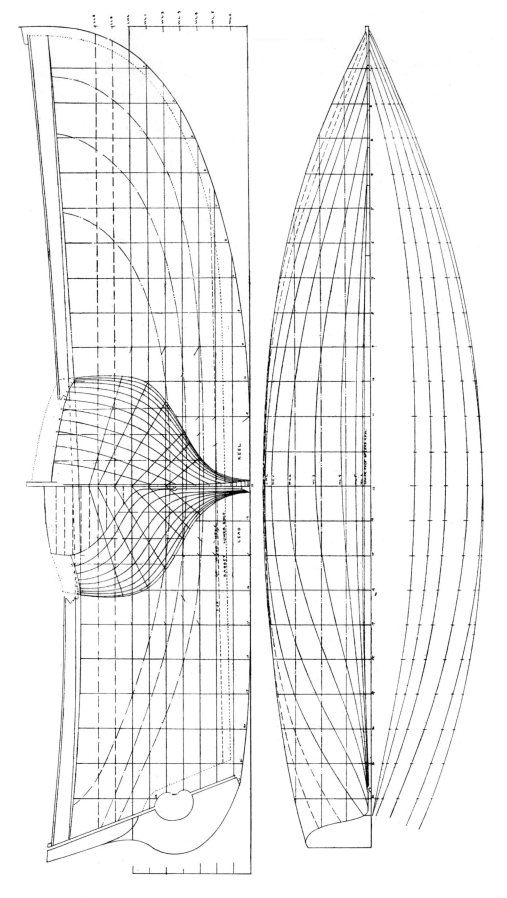

30-ton auxiliary cutter BETTY, designed for Mr C Hellyer and built by Messrs. Stow & Son at Shoreham.

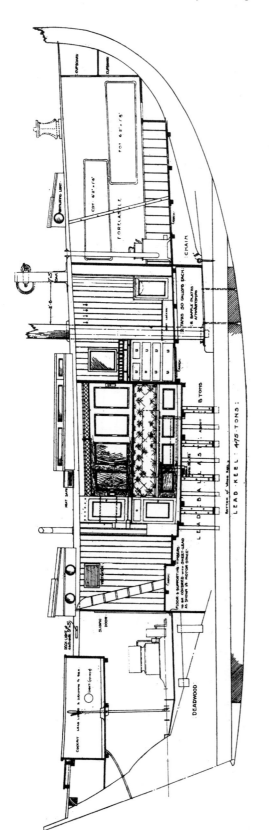

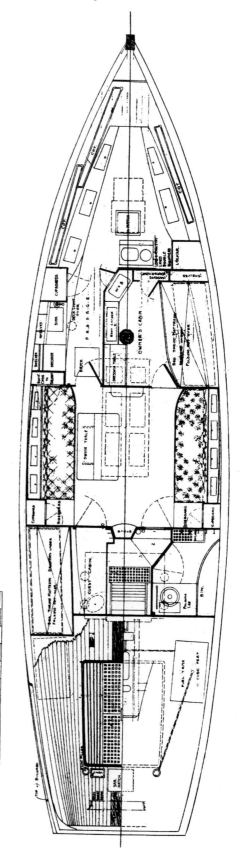

Auxiliary cutter BETTY.

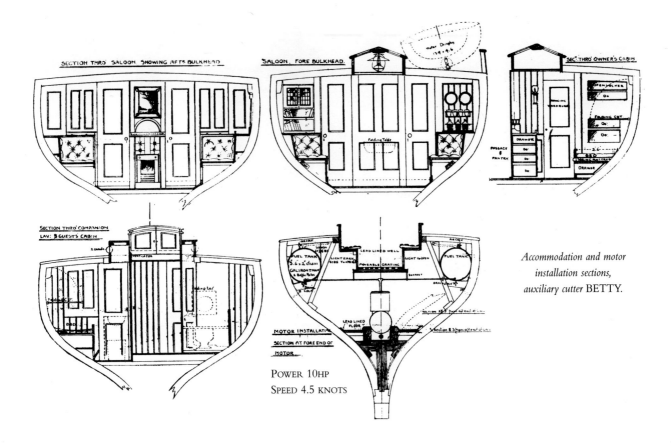

Accommodation and motor installation sections, auxiliary cutter BETTY.

POWER 10HP
SPEED 4.5 KNOTS

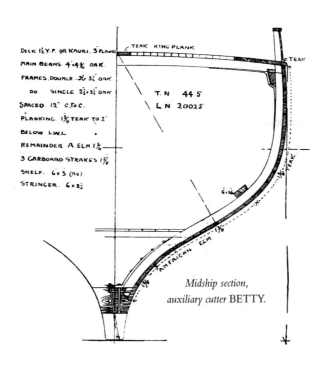

Midship section, auxiliary cutter BETTY.

Originally named BETTY, TALLY HO is seen here in her racing rig, while in the ownership of Lord Stalbridge in the 1920s. She carries a fidded topmast (compare with original sail plan, page 136), and her spinnaker is set on a long boom. Such booms could not be rigged with an effective downhaul and a crew member is seen at the weather shrouds lending his weight to help prevent the spar from lifting. Photograph: courtesy Beken of Cowes.

DESIRE 1909

LOA: 28ft 10in	LWL: 22ft 0in	Beam: 5ft 5in	Draught: 4ft 1in
Sail Area: 507sq.ft	Disp: 2.57 tons	Ballast keel: 3224 lbs	

Strange produced this design as a competition entry. The conditions called for a Six Metre racing yacht designed to comply with the International Rule and for building to the 'R' class at Lloyd's (which governed scantlings).
The Yachting Monthly judges commented:

Desire is too narrow and too long. Her bilge we do not like and we question if the boat could take advantage of it at any angle.

Although Strange designed few racing yachts, other than model yachts, it is obvious from his letters to yachting magazines that he had a very good grasp of the rating rules.

The winning entry was submitted by P C Crossley and designs by F C Morgan Giles and E P Hart were commended.

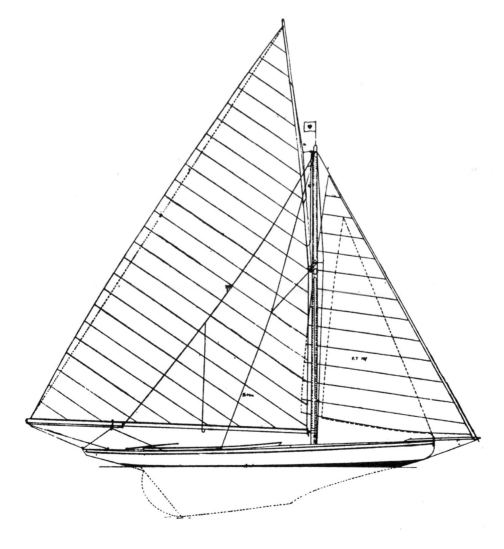

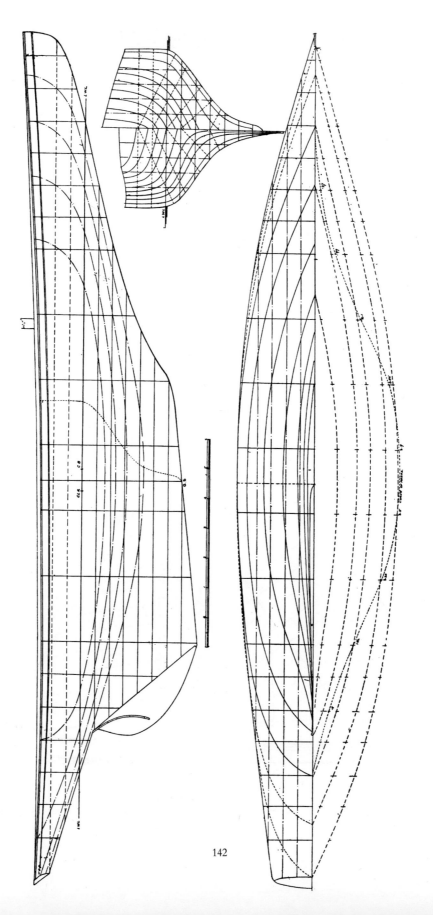

DESIRE

PUFFIN II (later named WOOZLE) 1909/1910   No.108

LOA: 44ft 7in	LWL: 39ft 6in	Beam: 9ft 2in	Draught: 3ft 6in
Sail Area: 340sq.ft		Ballast Keel: 14cwt	TM: 13 tons

This is believed to be the only motor cruiser that Strange ever designed. George C Waud, for whom she was built, was a member of both the Royal Yorkshire Yacht Club and the Scarborough Sailing Club. As Strange was also a prominent member of this latter club, it can safely be assumed they knew each other. Strange commenced work on the design in September 1909 and she was built in 1910 at Shipley, Yorkshire, by G E Ramsey, believed to be a firm of timber merchants.

The yacht was 'intended for day cruising and fishing on the North Sea', which explains her lack of accommodation. The narrow beam would ensure a reasonable speed (10 knots was expected), without the need for excessive horse-power or fuel consumption. Her 340sq. ft. of sail and alternative tiller steering would ensure she could make port in the event of mechanical failure.

Her first engine was a three-cylinder paraffin motor built by Fay and Bowen in the USA. This was replaced by a four-cylinder petrol engine in 1925. In 1932 she was fitted with another four-cylinder engine, this time a Parsons.

Although apparently based on the northeast coast while owned by Waud, later owners were from the Solent area. She apparently survived until 1957. She is listed in Lloyd's Register of Yachts for that year as 'No longer in existence'.

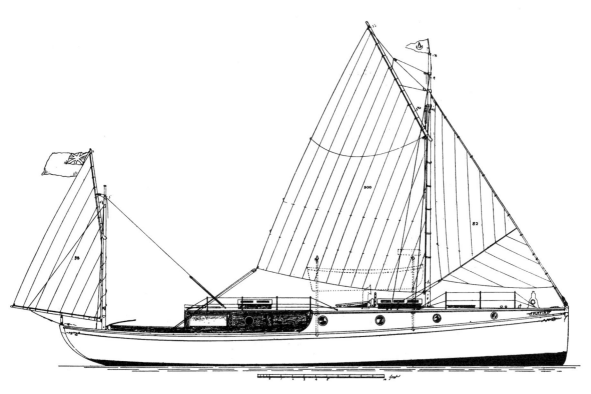

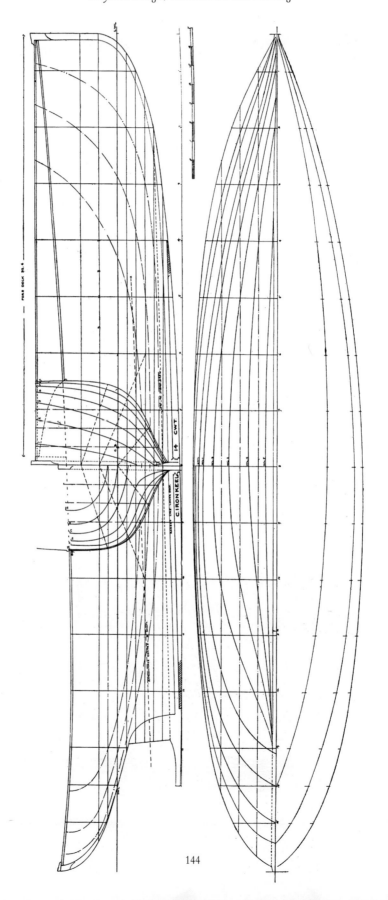

Motor cruising yacht PUFFIN II designed for Mr G C Waud and built by G E Ramsey of Shipley, Yorkshire.

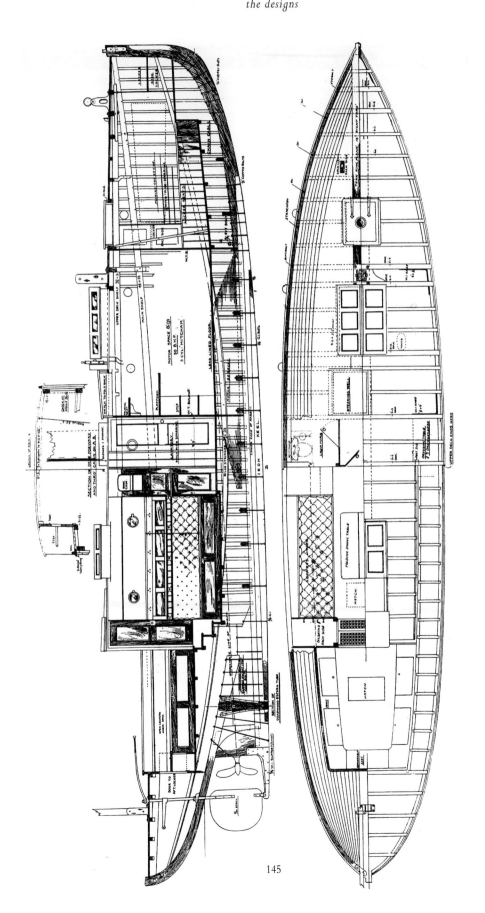

Motor cruiser PUFFIN II.

NORMA/SHULAH 1910   No.110

LOA: 25ft 5in	LWL: 20ft 0in	Beam: 7ft 2in	Draught: 3ft 5in
Sail Area: 365sq.ft	Disp: 3.35 tons	Ballast Keel: 1.33 tons	

When Strange designed this delightful little yawl for Norman Jones in March 1910, the drawings carried the name *Norma*. But when Jones had her built in 1914 by Bond of Rock Ferry, Merseyside, she was sloop-rigged and called *Shulah*.

Jones evidently changed his mind not only about the rig and the name, but about the boat itself – 18 months after the *Norma* design was prepared, Strange drew up a second design for Jones of a quite different type, initially known as *Gannet*, dated 1st October, 1911. A 27ft 8in canoe yawl, in the same mould as, but much larger than, *Cherub II* and *Tavie II*, *Gannet* was built and registered as *Psyche* in 1912 (see page 122 and JL p.115f.). Jones can have had only one, or at the most, two seasons in *Psyche* before reverting to the *Norma* design. *Psyche* was sold in 1913 to Captain G A E Clarke, who later commissioned Strange to design *The Jilt* for him (see page 201f.).

Norma was based at Hoylake on the tip of the Wirral peninsular, where her shallow draught will have been an advantage. Norman Jones kept her for about seven years. The next owner had only one season with her before she was acquired by Captain R D Briercliffe, a well-known yachtsman in the locality. Sometime in the early 1920s she was re-rigged as a yawl, presumably to the designed sail plan, but in 1924, Captain Briercliffe changed the rig back to sloop.

In January 1936 there was a report in *The Yachting Monthly* that *Norma* was being fitted with a bermudan-wishbone rig. This consisted of a bermudan mainsail and wishbone boom, set slightly higher than the ordinary boom and parallel to the waterline, the tack of the mainsail coming well down the mast, while the foot of the sail flowed up to the outer end of the boom. The wishbone pivoted on the mast in a manner similar to a gaff. Apparently this novel arrangement was successful and Briercliffe reported that *Shulah* was very easy to handle. He owned her for 18 years until she broke away from her moorings at Glyn Garth in the Menai Straits in a heavy SW gale late in 1940 and was lost.

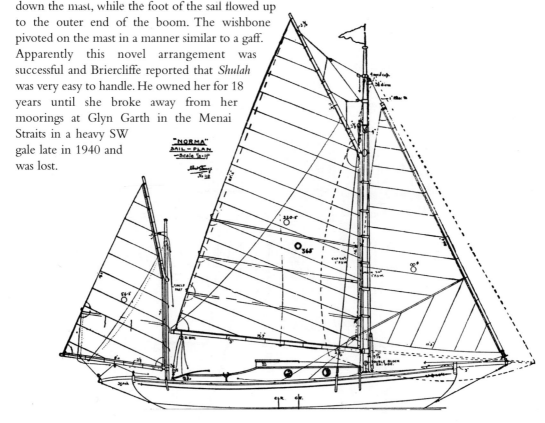

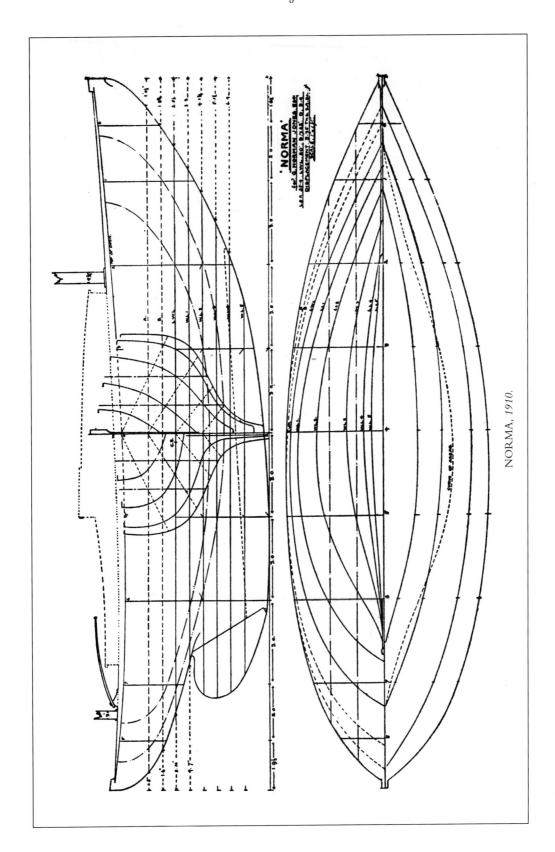

NORMA, 1910.

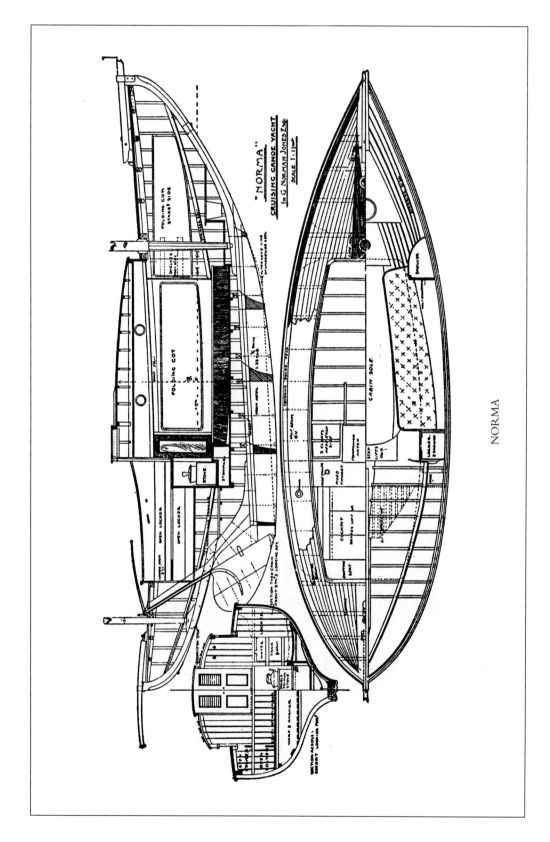

IMOGEN II 1910   No.116

LOA: 25ft 9in	LWL: 22ft 6in	Beam: 7ft 11in	Draught: 3ft 6in
Sail Area: 399sq.ft	Disp: 4.35 tons	Ballast Keel: 36cwt	TM: 6 tons

Imogen II is one of a handful of transom-stern yawls designed by Strange. Another is, of course, the original *Cherub III* design (see page 154f.). Both were drawn in 1910, *Imogen* in April, *Cherub* in October. By comparing the two, it is interesting to see how Strange has produced, in *Cherub III*, a much bigger boat on the same overall length. No doubt the recently completed design process for the one played a part in the development of the other. *Cherub III* has an extra inch or two on the waterline, beam, and draught, but the real difference is made by the fuller sections and the increase in freeboard – 3in more than *Imogen II* at the bow and 6in at the stern.

Imogen achieved some distinction under one of her later owners, Captain Otway Waller, in 1930. In October of that year, *The Yachting Monthly* carried an article under the title, 'A Device for Running at Last! An Invention Proven Practical on 1,600-mile Single-handed Ocean Cruise'.

> *A thoroughly practical device by which a small fore-and-aft rigged yacht can be made to run before wind and sea with no one at the helm has been invented and thoroughly tested by Captain Otway Waller who, in June to August last, sailed a 6-ton yawl single-handed from the Shannon to Las Palmas, Canary Islands.*
>
> *It has hitherto been generally acknowledged that a fore-and-aft rigged vessel will not run before the wind, as a model yacht can be made to do, with no one at the helm, even though she may be able to look after herself on any other point of sailing. Here and there has appeared a solitary exception, such as Captain Slocum's famous* Spray, *which was said to run before the wind unattended for days on end. But for the great majority of cruising yachts it may truthfully be acknowledged that they cannot be left to run before the wind and sea for long, even with only a squaresail set.*
>
> *Captain Waller's invention, which he described in an interview to a representative of* The Yachting Monthly *and expressed his wish that it should be published for the benefit of our readers, is both practical and relatively simple. It is strong, easily set and quickly stowed and with it his little yawl* Imogen, *22½ft on the waterline, ran 105, 107 and 106 miles on consecutive days without anyone touching the helm. It is not merely a 'light weather gadget' but proved well able to stand up to a strong blow.*
>
> *In fitting this device to a yacht the existing rig need not be altered. The device consists, briefly, of a pair of small spinnakers with booms, guys and roller gear. That is all. Looking at it one is tempted to ask: Why hadn't it been thought of before? It may well prove a boon to ocean-cruising men who adapt it to their yachts in future.*
>
> *The photograph shows Captain Waller's boat* Imogen *with the two running sails set. This was taken at the bottom of his garden in the heart of Ireland, a hundred miles up country from the mouth of the Shannon, a few days before he started on his lone passage to Las Palmas.*

The article goes on to describe the rigging of the booms in more detail and the lead of the sheets, through blocks on the quarters, to the tiller. Waller's spinnakers were 14ft on the foot and 120sq. ft each in area, together equalling the area of the mainsail. They were tacked down to the base of the mast and each was set on a Wykeham Martin furling gear. Nowadays, twin staysails are more commonly used, set to the stem head, which I suspect reduces the tendency to roll.

Waller encountered strong head winds when he set out and spent considerable periods hove-to.

> Imogen *was splendid hove-to, as dry as a bone … I found her just about as good as a small boat could be for the job, and her 'lack of draught' as some yachtsmen might describe it, eased her considerably in a sea way.*
>
> *You see I found she fore-reached rather fast when hove-to and was inclined to catch an advancing crest on her weather shoulder. The jar that it gave her shook the whole boat and I could hardly sleep. So I*

lashed the main boom to the weather mizen shrouds – the mainsail was close reefed – and with the staysail aweather and the helm hard down she made no headway at all but drifted to leeward bodily, leaving a smooth wake to windward for about 10ft or more. That wake seemed to break up the seas almost like oil and the crests usually spent themselves before they could reach her. She rode perfectly comfortable like this with her decks quite dry . . . Of course you want sea room to leeward and it may not be 'usual' to lie to like that, but it seemed to suit Imogen.

A bout of fever prevented Waller from continuing his voyage beyond Las Palmas and, later, his plans to circumnavigate were shelved. It is believed that *Imogen II* returned to the UK as deck cargo. For a time Waller taught astro-navigation in London and kept *Imogen II* until 1935. She then spent some years in Chichester Harbour, then moved with another owner to St Helier in Jersey. For the past 30 years or so, she has been based in the Torbay area of South Devon. In 1961, she was re-rigged as a bermudan yawl, but changed back to gaff in 1981 with a new set of spars.

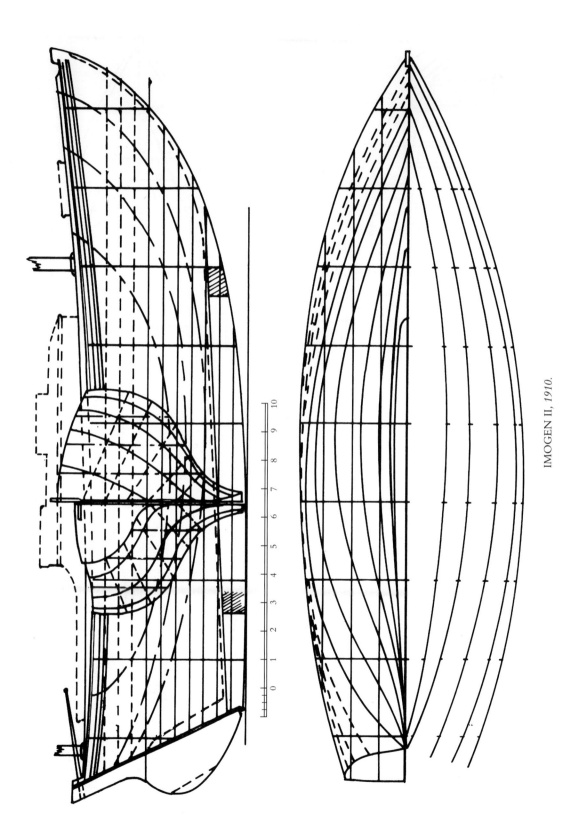

0 1 2 3 4 5 6 7 8 9 10

IMOGEN II, *1910.*

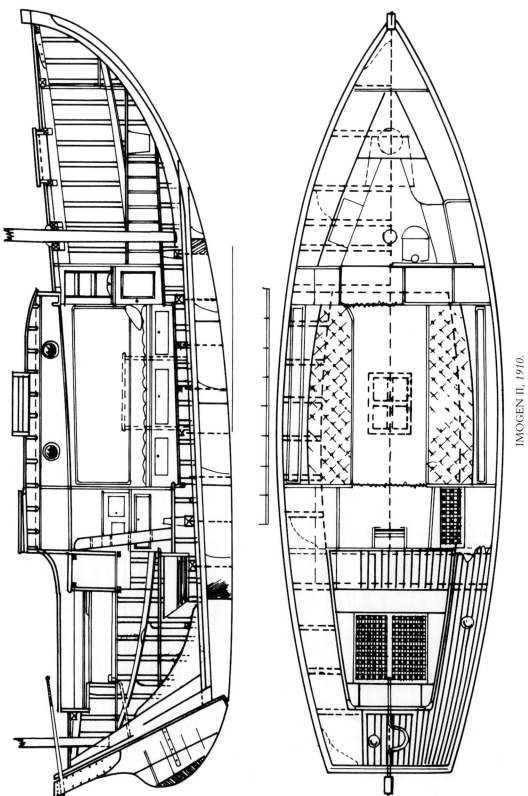

IMOGEN II, 1910.

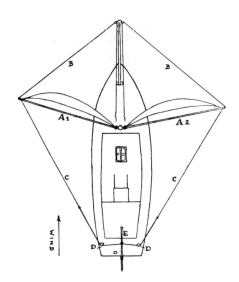

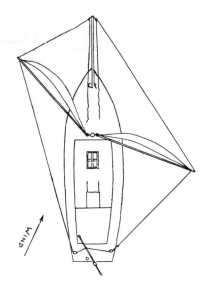

Above:
IMOGEN II *with the
running sails set,
at* **left**, *the sails rolled
up and the two booms
in place.*

CHERUB III Transom–stern Version 1910   No.112

LOA: 25ft 10in	LWL: 22ft 9in	Beam: 8ft 3in	Draught: 3ft 9in
Sail Area: 480sq.ft	Disp: 4.6 tons (approx)	Ballast Keel: 31.6cwt	

Gretta, Ariel, Galatea, Emerald

One of Strange's favourite cruising grounds was the west coast of Scotland. *Cherub III* was designed for his own use with this area in mind. Good freeboard made for a dry boat with interior volume sufficient to accommodate a sleeping cabin between the saloon and focsle, 'in order that a lady member of the crew could have somewhere to withdraw to when the society of the male members grew oppressive'. The light draught of 3ft 9in would allow the artist to bring the yacht into those shallow corners which afforded the best sketching.

Writing of the design in the Humber Yawl Club Yearbook of 1911, Strange explains how the boat came to be built with a canoe stern, rather than the transom shown here

> Cherub III *was originally designed with a square stern, but the plan was departed from in obedience to a general outcry on the part of many admirers of the canoe stern. This departure, having been successfully overcome from the financial side, has not been regretted by the owner and certainly adds much to the appearance and general efficiency of the yacht.*

REDWING,
ex-CHERUB III
Photograph:
L'Atelier, Douarnenez

Strange was well pleased with his new boat,

She turned out to be handy, extremely stiff and not at all slow regarded from a cruising point of view. Her light draught was not found detrimental to good windward work. In headroom, ease of handling, room on deck and in the cockpit, she fitted her crew's requirements capitally and her bold side and business-like look were much admired.

Gretta was built to the same design in Ireland in 1912. Strange amended the sail plan, increasing the sail area by 11% from 480sq. ft to 530sq. ft. He also re-drew the keel profile to run parallel to the LWL from about amidships to the heel of the rudder. This would have simplified taking the ground.

In 1937 another sistership, *Emerald* was built in Deganwy, North Wales, by Williams and Parkinson. Williams recorded that he went to a lot of trouble with the lofting to get the shape. *Emerald*, like *Cherub III*, was built with the ladies' cabin. Both she and *Cherub III*, now named *Redwing*, have been sailed and cruised widely by recent owners. *Gretta* was less fortunate. In 1954, while on passage in bad weather from Holyhead to Fleetwood, she foundered after hitting the stone revetment on the edge of the deep-water channel in the Mersey. Her crew, including dog, were rescued, but the yacht was lost.

However, the original transom-stern design had not gone unnoticed – in 1925 *Ariel* had been built at the yard of Anderson, Rigden and Perkins, at Whitstable, Kent. The keel and garboards were altered slightly, to give 6in more draught and to simplify construction. Rigged as a gaff cutter, she was fitted out with conventional two-berth saloon and focsle under a straight-sided cabin. The deck aft was run inside the cockpit coamings with no separate seats.

She proved a successful boat, taking her second owner, J B Kirkpatrick, on an extended cruise to Spain and back in 1928. An account of this cruise appeared in *The Yachting Monthly* and also in Kirkpatrick's book, *Little Ship Wanderings*.

In 1930, when *Ariel's* builders were short of orders, her building moulds were again set up and a sistership built on spec. Cutter-rigged like her sister, this boat was named *Galatea* and is still based in the Whitstable area. She has been owned for some 25 years by Richard Blomfield, a long-standing member of the Albert Strange Association. In 1995 he had *Galatea* re-rigged as a yawl, following closely Strange's sail plan for the original *Cherub III* design.

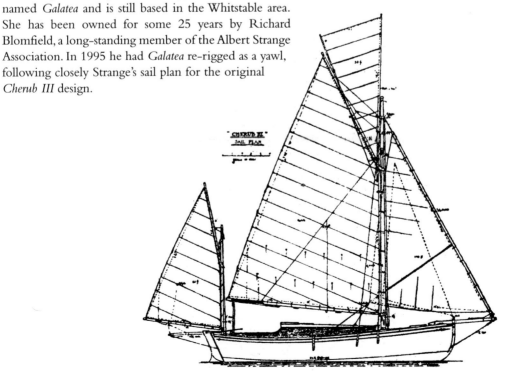

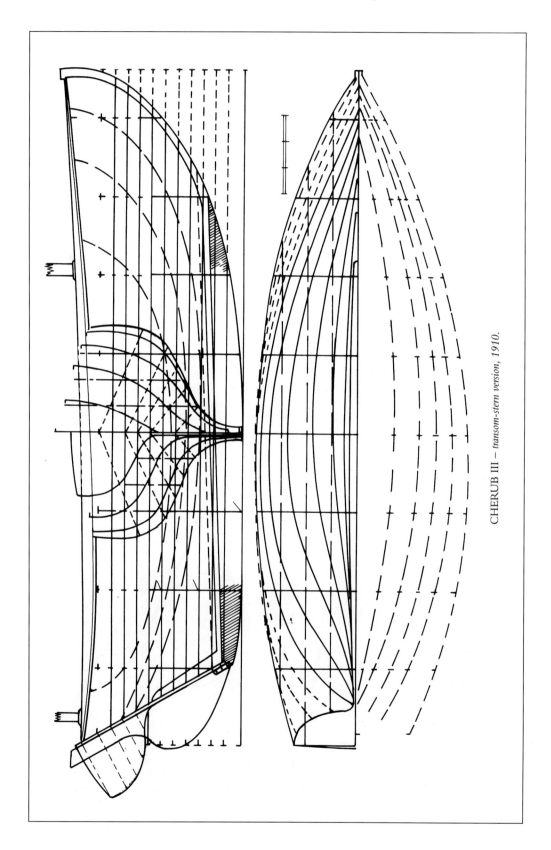

CHERUB III – *transom-stern version, 1910.*

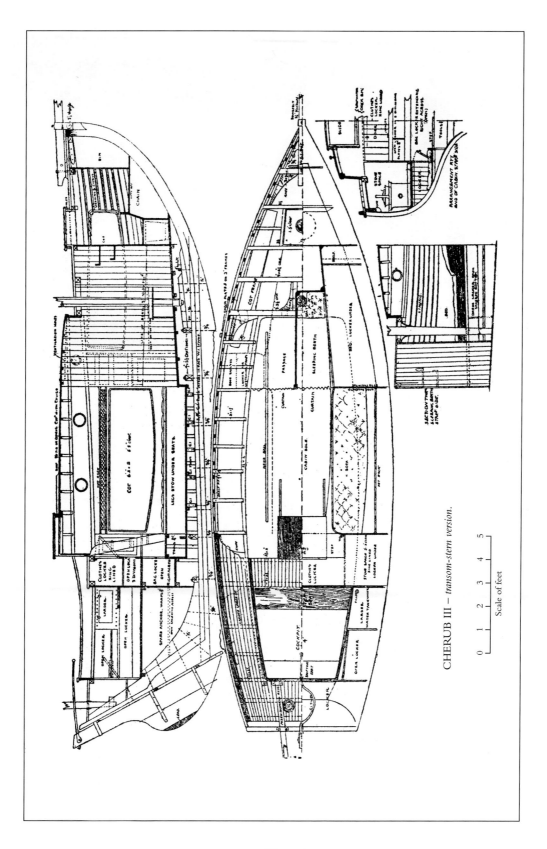

CHERUB III – *transom-stern version.*

Scale of feet

0 1 2 3 4 5

ARIEL, *owned by Jim, Jo and
Peter Maynard of Maldon, Essex.
Main photograph: Den Phillips.
Photograph, left: Peter Benstead.*

Below: EMERALD, *passing the Point of Ayre in a NE 5-6,
May 1987. At the time she was owned by Joe Pennington and
was taking part in the annual race around the Isle of Man.
Her sister-ship,* REDWING, *ex-*CHERUB III,
*was not far ahead.
Photograph: Rick Tomlinson.*

FIREFLY 1922

LOA: 34ft 9in	LWL: 24ft 0in	Beam: 7ft 9in	Draught: 3ft 9in
Sail Area: 500sq.ft	Disp: 6 tons (approx)		TM: 7 tons

F*irefly* is something of a hybrid. In the short article reporting her launch in *The Yachting Monthly* for December 1922, it is stated that her owner, Mr A J R Lamb of the Royal Norfolk and Suffolk Yacht Club

> *... wanted a cruiser which would be equally suitable for use on the Norfolk Broads and for sea work and* Firefly *was built to a design of the late Albert Strange – the designer of so many charming little cruisers – modified by Dr Harrison Butler. The modifications consisted mainly in the conversion from cutter to yawl rig and the reduction of draught to 3ft 6in. The latter modification is, of course a compromise, but on the waterline length of 24ft, should be sufficient for the requirements of sea work, while not being prohibitive for Broads sailing.*

The original Strange design has not been positively identified, but it seems likely that it was design *No. 119*, the 7-ton Auxiliary Sloop (see page 163 and JL Fig. 48-50), despite the above reference to the change being from a cutter rig. The lack of a bowsprit (for sailing on the narrow and crowded waters of the Broads), as well as the engine installation, could well have been aspects of the Auxiliary Sloop which attracted Lamb's notice. The auxiliary centreboard yawl which appears on pages 62, 63 and 95 also shows remarkable similarities in profile and accommodation, though her much greater beam and longer waterline preclude her from being the actual design that was modified. A comparison of the two is nevertheless instructive.

Firefly was built to unusually high standards for a yacht of her size by J W Brook at Lowestoft. Her first owner sold her after only three seasons due to ill health, but bought her back again in 1932. She only changed hands again on his death in 1934, when she was bought by Henry, later Sir Henry, Clay. She has remained in the same family for the 65 years since then.

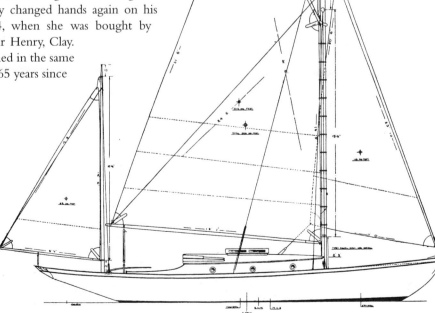

One or two early cruises in *Firefly* brought Clay south to the Harwich rivers, where he would sometimes meet the author, Arthur Ransome, who had a cottage and kept his boat at Pin Mill. Ransome was evidently impressed by the way in which Henry's wife, Gladys, coped with four children on such a small boat for, in 1937, he dedicated his children's novel *We Didn't Mean To Go To Sea* to Mrs Henry Clay.

In 1935 a bowsprit was added to reduce weather helm and, at the same time, the long forehatch, which allowed the heel of the counter-balanced mast to swing up through the deck, was converted to a conventional hatch. Apart from these alterations, *Firefly* remains virtually as built. Her yellow pine deck was replaced with teak in 1985, following the original scheme which is accurately shown in the accompanying drawing.

In 1994 her engine was removed for access to keel bolts and it was decided to sail her that season without auxiliary power. The propeller aperture in the rudder was blanked off and faired to the stern post, which produced a dramatic improvement in her performance and handling. This result draws attention to the fact that the performance of old yachts, originally built without engines, tends to be unfairly assessed: most have had engines fitted and the changes to the rudder configuration, the extra weight, and the drag caused by the propeller, can often severely spoil or even ruin their performance.

With her large cockpit and absence of bridge deck, *Firefly* has never been asked to voyage far afield, although she has visited the Baltic on three occasions, via the Kiel Canal. The first of these voyages was in 1937. During the North Sea crossing, the dinghy was fitted with a tightly sheeted cover and towed astern by a wire rope painter. Between Nordeney and the Elbe, the painter parted during the night and the dinghy was lost. Two days later, *Firefly's* crew were hailed in the Kiel Canal by a Dutch freighter which had retrieved the dinghy and hoisted it aboard in the vicinity of the Elbe No 2 light vessel. In the Westerly Force 7, the little boat had drifted 30 miles, crossing the very sands made famous by Erskine Childers in his novel, *The Riddle of the Sands*. The crew of the freighter asked for and were gratefully given the sum of £2 for their trouble.

FIREFLY *at the time of her launch.*

FIREFLY

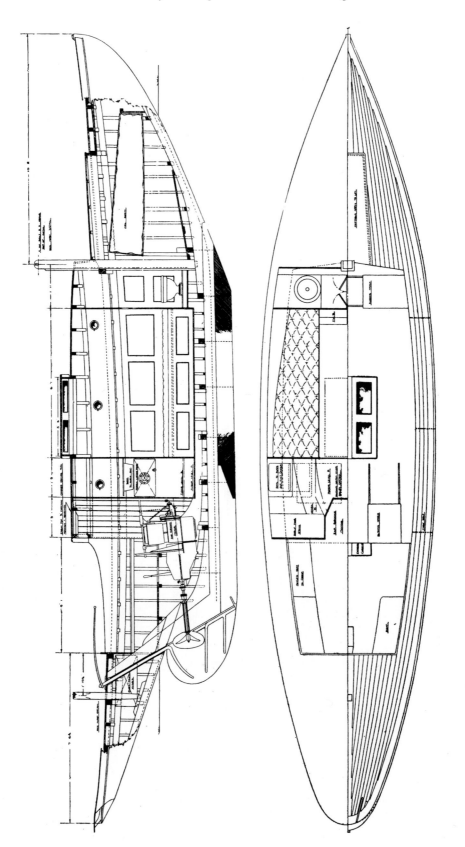

Auxiliary cruiser, FIREFLY.

GAFF SLOOP/YAWL 1911   No.119

LOA: 34ft 8in	LWL: 24ft 0in	Beam: 7ft 11in	Draught: 4ft 3in
Sail Area: 567sq.ft (Sloop), 570sq.ft (Yawl)		Disp: 5.1 tons	TM: 7 tons

Shown here is one of the working drawings for a design which became the 7-ton Auxiliary Sloop in its finished form, published in *The Yachting Monthly* in 1911. The drawing is from the collection at Mystic Seaport Museum and has been reduced from a scale of ⅜in to 1ft. The design was for 'an East Coast yachtsman who will cruise without a paid hand. Hence the small and easily worked sail plan'.

This drawing shows that a yawl rig was evidently under consideration at some stage, as well as demonstrating the amount of arithmetic involved in working out the areas and centres of effort of the sail plan.

In her yawl rig, she bears a remarkable resemblance to *Firefly*. Her principal dimensions and many of the details of her scantlings and specification are virtually identical and it is tempting to see No. 119 as the design on which *Firefly* was based.

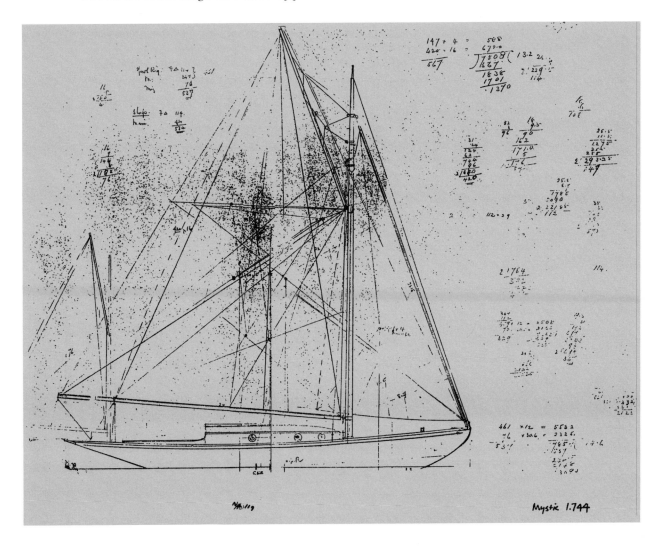

AFREET 1911

LOA: 30ft 0in	LWL: 20ft 0in	Beam: 6ft 11in	Draught: 4ft 3in (Lead)
			4ft 4in (Iron)
Sail Area: 440sq.ft	Disp: 2.45 tons	Ballast – Lead keel: 2,502 lbs	
		Iron keel: 2,558 lbs	

This little boat was designed for the 12th competition in *The Yachting Monthly* in 1911. The competitions were popular and covered a wide variety of types. Professional as well as amateur designers frequently participated. The prize was usually £10 – a substantial sum of money when it is considered that £150 would buy a brand new 30ft day boat.

Printed below is an extract from the rules for this competition as they appeared in the October 1911 issue of the magazine.

We offer a prize of £10 for the best design for a small restricted racing yacht conforming to the following conditions:

The boat is to be primarily for racing, of reasonably stout construction, not necessarily with accommodation and with an average cost price of £150.

No dimensions are suggested, but it may be borne in mind that the proposed boat is for universal use and must be capable of cheap transit by land or water. The constructional design must be simple in order to reduce first cost. A fast boat is wanted and one which can be efficiently raced by two and comfortably sailed by one.

It may be suggested, further, that good initial stability is desirable, a model capable of easy beaching and lines and sail plan which will ensure speed and handiness without undue sacrifice of seaworthiness. There must be no centreboard and no inside ballast.

Each design must be accompanied by a brief specification and a suggestion of restrictions should be so drawn up that some latitude is left to the designer. At the same time the owner's interests must be safeguarded and the measurement rules must be sufficiently stringent to prevent his being readily outbuilt.

Twenty-three entries were submitted and the result was published in December 1911. The prize went to Robert Cole, a friend of Strange's and fellow member of the Humber Yawl Club. Six designs accompanied the judges' critique, another followed in January 1912 and three more, including *Afreet*, appeared in the February number.

The judges' comment on Strange's design was:

Afreet, although a good boat, is too large. The lines are excellent, if hardly as powerful as the dimensions warrant. £150 would be a low quotation for her.

This competition gave rise to some interesting correspondence concerning the cost of building. It came from Strange and E P Hart, both of whose designs had been criticised for being too expensive to build. There were no 'sour grapes' involved. Strange, as so often, had the interests of others in view when he wrote, in his usual engaging style...

Most of us would like to go to a Bond Street tailor if we could afford to, but doubtless most of us have to be content to walk the earth clothed in garments built by local artists, who, though they do not charge so much as their Metropolitan brethren for reputation and 'cut', use exactly the same cloth and produce very much the same result.

Please, Sir, do not scare off the young beginner by making him think that none but two or three extremely high-class builders can produce a boat fit to be seen alive in, because, in the first place, this is not true and, in the second place, it is extremely injurious to the very many smaller firms of yacht builders all over the country who are turning out excellent work at reasonable prices.

E P Hart pointed out that he had placed the plans of his design before a well-known firm, paying

trade union rates, who had agreed to build a single boat for £150. The editor replied that £150 had been intended as a maximum and that a design able to be built for less, weighed in the designer's favour.

In his letter, Strange had cited as an example of a boat almost identical in its dimensions to Cole's winning design, the Sea View OD Class, built in the Isle of Wight in 1908: LOA 25ft; LWL 17ft; Beam 6ft; Draught 3ft 7in; Displacement 34cwt; Lead 15cwt; Sail Area 300sq.ft. The cost of this boat was £70. If the cost of Cole's boat was over double this amount, the customer was not getting value for money. The class, known as the Sea View Mermaid was designed by G U Laws.

It is not possible to say, at this distance in time, what influence the designs in this competition and the correspondence that followed may have had, but it is worth recording that during the following year, G U Laws was commissioned to design a new one-design class boat for the Royal Corithian Yacht Club at Burnham-on-Crouch in Essex. The class later became known as the East Coast One Design, the first batch of which was ready for the 1913 season, and in December of that year, *The Yachting Monthly* carried a short report, together with the lines (see page 167).

Laws' design is in all respects a 'big sister' of the Sea View Mermaid. Her length overall and on the waterline are identical with *Afreet*, but Laws had a fixed price limit of £120, which was no doubt the reason for adopting the quicker and cheaper fin keel construction. The hull was planked upside down, with no hollow curves, and the keel assembly bolted to it through the floors after the hull was inverted. Nevertheless, W King & Son, who built them, evidently found it difficult to build within the price, since only the first boat, *Chittabob IV*, had plank fastenings clenched over roves; in subsequent boats, the copper nails were only turned. They raced as a class into the 1970s and, with their light construction, it is remarkable that several are still sailing today.

A comparison with *Afreet* shows Laws' design has 6in less beam, 3in less draught and 270lb less displacement, with 10sq.ft more sail area, but no spinnaker. They are notably fast in light weather, but I suspect *Afreet* would have been the better heavy-weather boat.

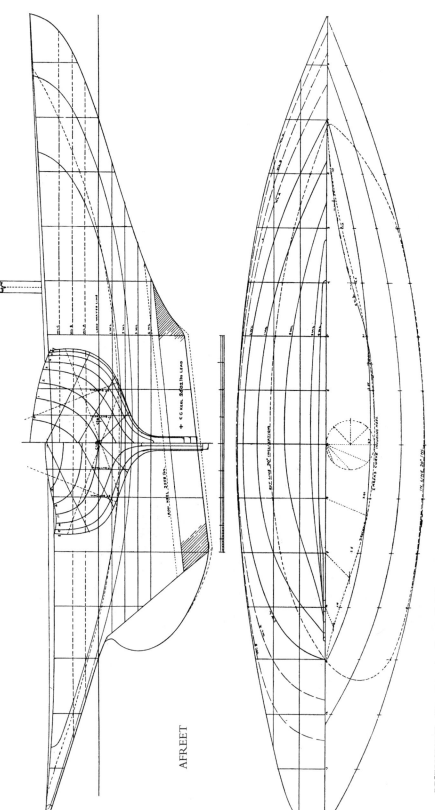

AFREET

RESTRICTIONS.—The extreme length over-all shall not exceed 30ft.
The length on the waterline shall not exceed 20ft., measured without crew on board.
The external beam to outside of planking shall not exceed 7ft. or be less than 6ft. 6ins.
The extreme draught shall not exceed 4ft. 6ins., or be less than 4ft.
The minimum depth inside the hull immediately abaft the mast, measured from the underside of deck plank to the inside of skin planking at rabbet, shall not be less than 3ft. 6ins.
The total area of fore and aft canvas shall not exceed 440 sq. ft.   The area of mainsail not to exceed ·75 of total sail.   The sails may be shaped and disposed as required by the owner,
    but must be at least in two pieces.
A spinnaker of 130 sq. ft. shall be allowed in addition to the fixed area of fore and aft canvas (?).   Actual area of sails to be measured.
The weight of the lead or iron keel shall not be less than 2,240lbs.   The weight certificate for keel to be signed by designer, owner, and builder.   No movable ballast of any kind
Any internal ballast required for trimming is to be securely and permanently fastened to the main keel to the satisfaction of the committee of inspection.
    to be allowed under any circumstances.
A length of at least 7ft. of the underside of the lead keel shall be a perfectly straight line in profile.
Any alteration in hull, spars, or sails required by the owner must be notified, with all particulars, to the committee of inspection, who will give a certificate of clearance after inspecting
    the alterations when completed to their satisfaction.
The total area of all deck openings, whether covered with hatches or not, shall not exceed 28 sq. ft., nor be less than 18 sq. ft.
The waterway shall in no case be less than 18ins. in width clear of coamings.
No hollow spars and no bamboo spars allowed.

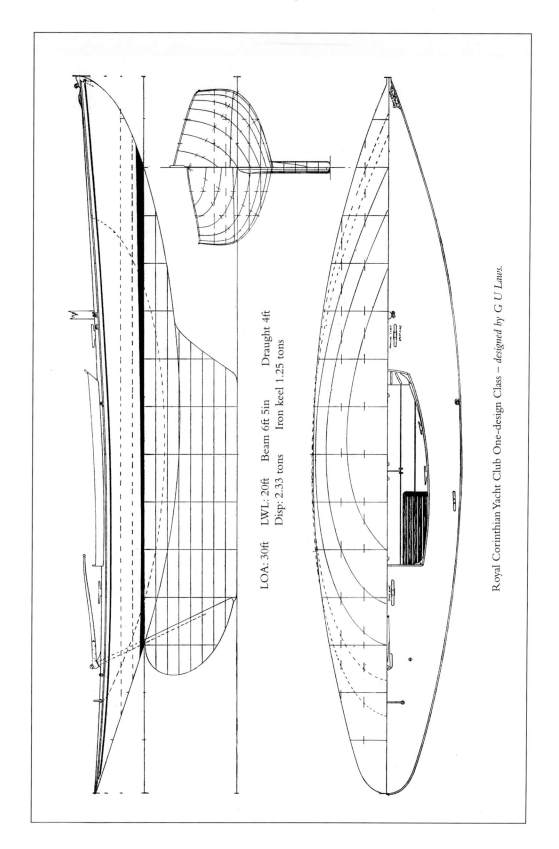

LOA: 30ft   LWL: 20ft   Beam 6ft 5in   Draught 4ft

Disp: 2.33 tons   Iron keel 1.25 tons

Royal Corinthian Yacht Club One-design Class – *designed by G U Laws.*

FLAPPER 1912

LOA: 24ft 2in	LWL: 17ft 0in	Beam: 5ft 7in	Draught: 3ft 2in
Sail Area: 265sq.ft	Disp:1.35 tons	Ballast Keel: 1,430 lbs	

Albert Strange submitted this design in December 1912 as his entry in *The Yachting Monthly* Design Competition No. 15.

Competitors were asked to design a small racing boat of 27cwts displacement. No other restrictions were imposed, but entries had to be of a type that could be easily driven and handled. Preference would be given to a cheap boat. Although racing was the primary object, there were other considerations, and the fastest boat would not find most favour, but rather, one which promised speed with comfort and safety. Centreboards were barred. A deep-bodied hull was not sought, but neither was an over-shallow one.

Apparently a large number of entries was received and the editor decided that, due to ever increasing demands on space, only the winning entries would be published.

The competition was won by Norman E Dallimore's design *Scud*, with *Flapper* being one of those commended.

Although Strange had produced a charming design, the judges felt that, with the reverse curve in her sections she would be more expensive to build than the winning design which had a simple fin keel. It is interesting to note the similarities between *Flapper* and *Afreet*, page 164f., the racing boat designed by Strange for *The Yachting Monthly* competition of the previous year and similarly criticised on grounds of expense.

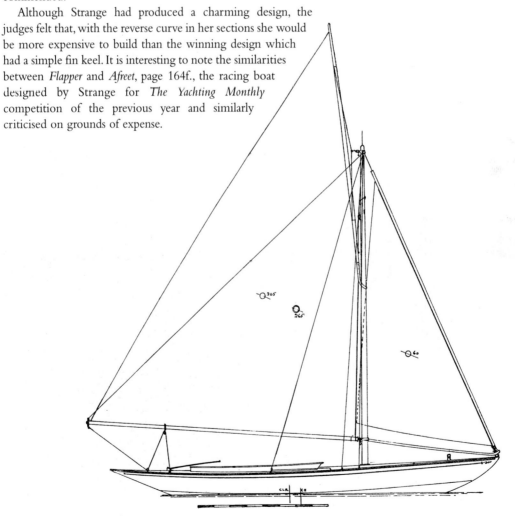

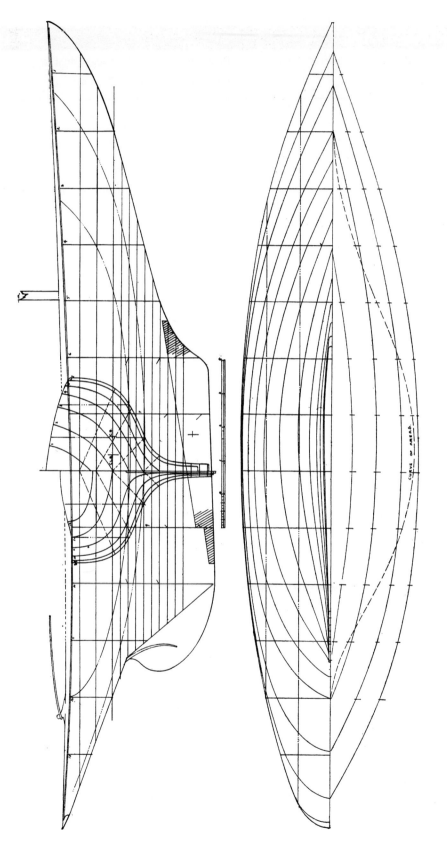

FLAPPER

CURVE OF AREAS

MOTH II 1912   No. 126

| LOA: 28ft 5in | LWL: 21ft 0in | Beam: 7ft 6½in | Draught: 4ft 1in |
| Sail Area: 410sq.ft | | Ballast Keel: 33 cwt | TM: 5 tons |

Albert Strange completed the design of *Moth II* early in 1912 for Harry H May and she was built by Luke of Hamble later that year. The first *Moth*, a four-ton sloop, had been designed and built in 1905 by Luke and was owned by May in partnership with Walter Beaumont. In 1908, Beaumont had Strange design the 32ft. canoe yawl *Hawk Moth* for him (see JL fig. 22-24). She was built by Dickie of Tarbert and was based there while Beaumont owned her. It is not known what influence this yacht may have had on Harry May's choice of design for his new boat, but *Moth II* was of similar size, though with greater interior volume.

May only sailed *Moth II* for a couple of seasons in the Plymouth area before putting her on the market. She was priced at £300 and the advertisement stated that she had cost £440 to build.

She left the West Country in 1922 when a new owner sailed her up to Conway, North Wales. This gentleman wrote that *Moth II* 'has a cabin that would do credit to many a 10-tonner. Her late owner, having been built to a mould of heroic size, had insisted on much space everywhere'. This is reflected in the full-bodied midsection and slightly more generous beam and freeboard than Strange usually gave to his designs in this style.

One of her subsequent owners obviously felt she was his ideal boat as he kept her for more than 20 years. She remained based on the northwest coast until 1952, when a terse entry in Lloyd's Register of Yachts for that year recorded 'Sold and reported wrecked'.

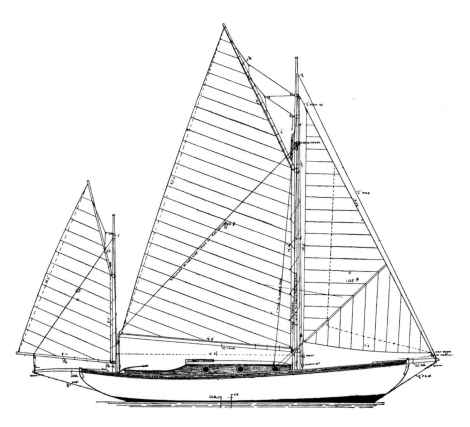

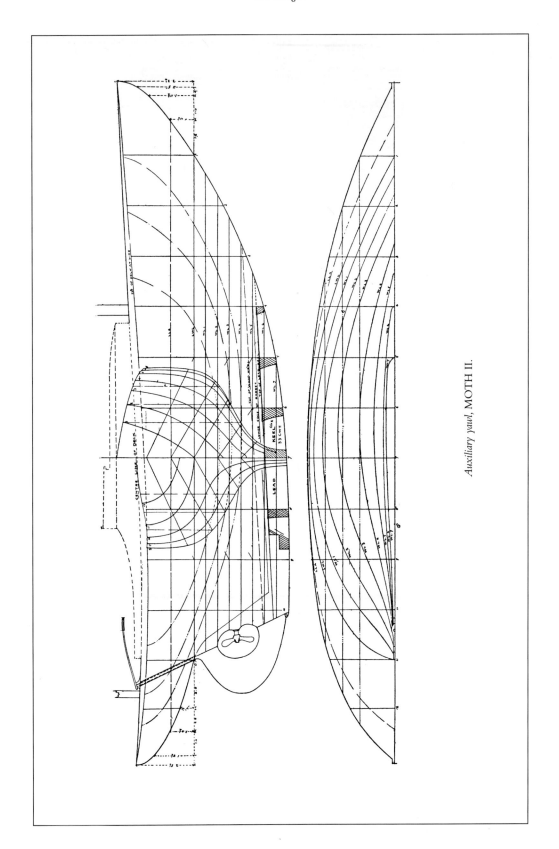

Auxiliary yawl, MOTH II.

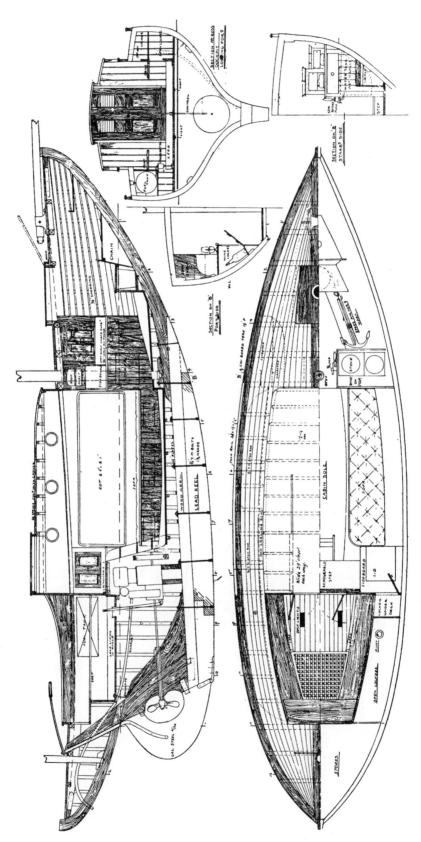

Auxiliary yawl MOTH II.

MOTH II

8½-ton Yawl NIRVANA OF ARKLOW 1914/1925   No. 144

LOA: 35ft 6in	LWL: 27ft 0in	Beam: 8ft 8in	Draught: 5ft 0in
Sail Area: 654sq.ft		Ballast Keel: 2.9 tons	TM: 8½ tons

The design of this yawl was commissioned by Lieut. Noel Constant and the finished drawings are dated December 1914. Presumably Constant's building plans were shelved for the duration of the Great War, after which his circumstances or inclinations changed. At any rate, he is listed in Lloyd's Register as the first owner of the 19-ton, 43ft. schooner *Lucette*, designed by Norman Dallimore and built in 1922 by Wm. King & Sons at Burnham-on-Crouch. (This yacht later achieved fame in the early 1970s when she was sunk by a killer whale in the Pacific while under the ownership of Dougal Robertson, who described his experiences in the best seller, *Survive the Savage Sea*.)

It was Strange's death in 1917 which prompted Constant to send the drawings of the yawl to *The Yachting Monthly*, in which they were duly published, in June 1918. This was fortunate, since it is unlikely that they would otherwise have come to the notice of A W Mooney from Dublin, who had *Nirvana of Arklow* built to the lines in 1925. Although he wanted a 'true single-hander', Mooney evidently considered Strange's sail plan not only 'snug', as the editor of *The Yachting Monthly* had commented, but *too* snug, for he drew a taller rig which increased the total sail area to about 700sq.ft.

Mooney also records, in an article in *The Yachting Monthly* in April 1929, how he re-drew the accommodation. By lengthening the coachroof, he managed to work a two-berth sleeping cabin between the saloon and the foscle. Having sailed a 400-mile passage in the boat, the writer can vouch for the excellence of Mooney's arrangement, which provides berths well away from the business of the cook and navigator, while retaining a foscle of useful size for the heads, an occasional berth, and good stowage.

The lines of the yacht as built appear to be as Strange drew them, but it is interesting to note that the propeller aperture has been cut entirely from the after deadwood. The efficiency and integrity of the rudder is thus much improved.

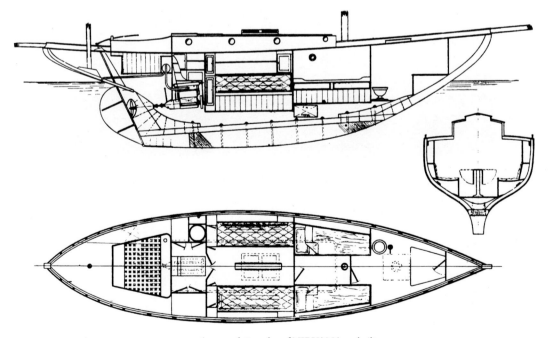

Accommodation plan of NIRVANA *as built.*

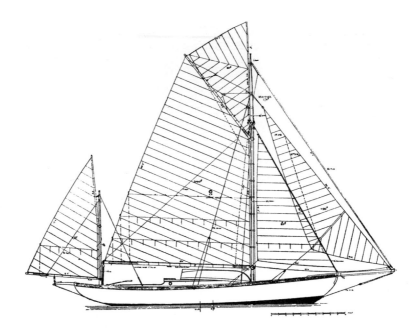

Nirvana was a great success, cruising extensively on the Irish and Scottish Coasts between Valentia in the southwest and Stornoway in the Outer Hebrides. In 1932, Mooney undertook a cruise with his family which took in many of the ports of the north and south coasts of Cornwall and Devon. On the outward leg, fetching down the East Coast of Ireland in a thick fog, he relates an incident involving the Tuskar Rock, the position of which he particularly wanted to be sure of, as it was his departure point for the long crossing to Land's End.

> *We could hear the five-minute fog explosion on the Tuskar and, wishing to see the rock if possible, as it is quite steep-to on its northern side, steered straight for it. Unfortunately the navigation was too good, for in one of the intervals between the explosions the rock loomed up suddenly, apparently almost overhead and before my crew, who was steering, could realise it, there was a jolt and there we were piled up. At that moment the gun went off directly overhead. I rushed on deck, saw deep water on both sides of us and realised we had crossed the boatslip to the rock and in a moment had the headsails down and soon* Nirvana *glided off into deep water.*

Mooney wrote that *Nirvana* had been 'such a success and so easy to handle by myself that I am seriously considering something larger, of at least 16 tons, still a single-hander'. In the event, a serious illness in 1933 forced the sale of *Nirvana*, and when Mooney had the 41ft ketch *Aideen* built in 1935, to the design of Fred Shepherd, her interior was laid out to accommodate a paid hand.

Nirvana continued to cruise in the same area under the ownership of Mrs. Ethel Crimmins who sailed with a paid hand, and several times before the Second World War won cups of the Irish Cruising Club and the Cruising Association.

Nirvana's present owners, Peter and Nancy Clay, who bought her in 1988, removed the unsightly doghouse which had been added during the 1950s, and slightly shortened the coachroof to create a new cockpit with curved coamings very similar to Strange's original design for Noel Constant. The offset companionway hatch has also been re-instated. It is interesting to note that these minor alterations have constituted such a practical improvement in the use of the cockpit and the galley, exemplifying Strange's skilful use of space in small yachts.

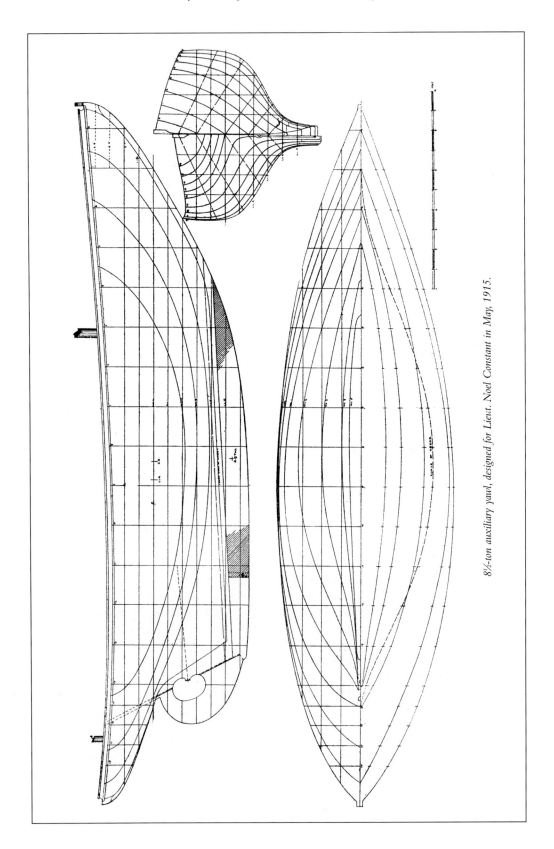

8½-ton auxiliary yawl, designed for Lieut. Noel Constant in May, 1915.

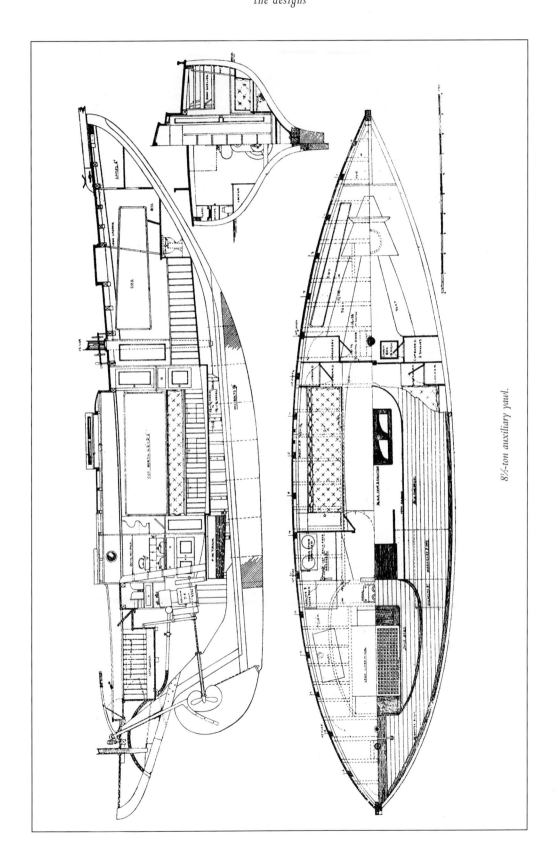

8½-ton auxiliary yawl.

NIRVANA OF ARKLOW

Owners Peter and Nancy Clay have had the doghouse
removed and the original line of the coachroof restored
as part of an extensive programme of renovation.
Below: on the River Blackwater, Essex, in 1993.
Photograph: Den Phillips.
Left: with a new suit of sails including gaff mizzen.
Photograph: Jamie Clay.

SINGLE-HANDED CRUISER 1910   No. 114

LOA: 31ft 7in	LWL: 29ft 0in	Beam: 10ft 0in	Draught: 5ft 0in
Sail Area: 593sq.ft	Disp: 10.1 tons	Ballast keel: 2.83 tons	Internal: 1.4 tons

This design was evidently commissioned by Claud Worth, from whose famous book, *Yacht Cruising*, the drawings were taken. The design is dated March 1910 and the first edition of Worth's book was published in November of the same year. Three years later, the keel of *Tern III* was laid – she was to be the 53ft cutter which Worth states he had been planning for several years and which Strange drew up for him. All of which would appear to suggest that Design No. 114 was commissioned as an exercise for the purpose of illustrating Worth's book, rather than with any intention to build. Worth's introductory remarks on the design seem to confirm this.

DESIGN FOR A SINGLE-HANDED CRUISER
The object of the design is to produce a boat suitable for all waters and weathers; for weekend sailing or for an ocean voyage. She would have ample accommodation for two or three people and in home cruising would be well within the capacity of a single hand.

The chapter on single-handed cruising was expanded in each subsequent edition of *Yacht Cruising*, but Worth retained Strange's design upon which to base his critique. The following is from the fourth edition, 1934.

For the man who is able to spend several months afloat, the design of the late Mr. Albert Strange is an admirable compromise. She would be under-canvassed, but the sails are large enough for a single hand and she would get along well in strong winds. Her motion would be rather too quick in a short sea, but she would be almost incapable of shipping a heavy sea. She would lie-to well under short canvas and would be fit to go anywhere. The idea of the round stern was to enable her, in case of necessity, to ride by the stern to a sea-anchor. The helm would then be lashed amidships and the boom lowered into an iron crutch about two feet high and there would be no gear to chafe the warp. I would prefer a small transome [sic]; also rather more rake to the keel. Next to seaworthiness, the main feature of the design is accommodation; she would be quite a floating home for a lone hand or for a man and his wife. Tara was built to this design and the owner speaks very highly of her.

(*Tara* does not appear in either British or American yacht registers and has so far not been traced.)

The work of Colin Archer had achieved considerable renown in Britain by this time and it is not surprising that Strange, like many others, was keen to try his hand at a design in the 'Norwegian' style. (He may indeed have had a more specific connection in Norway for, in November 1909, he had produced a 'sketch plan of an auxiliary motor fishing boat for the Lofoten Fisheries' for S A Fangen, showing a 32ft 6in commercial vessel.)

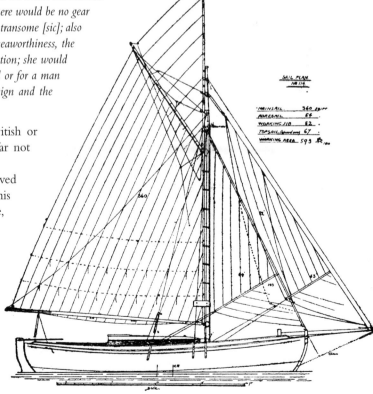

SAIL PLAN
Nº 114

MAINSAIL ... 360
MIZZEN ...
WORKING JIB ...
TOPSAIL ... 67
WORKING AREA 593

The interest of Design No. 114 lies in the way Strange has put his own mark on the type. Never a proponent of great beam, he has drawn a rather more slender hull than would be usual. The forebody is considerably fine, as indicated by the waterlines. The sections of the afterbody show a quite pronounced 'wine-glass' shape, a correspondingly more gentle rise to the buttock lines and a hollow run, where the Archer type has very slack, almost 'vee' sections and steeply rising buttocks. In consequence, Strange has lost the striking symmetry in the waterlines characteristic of Archer's hulls and moved the centre of buoyancy aft.

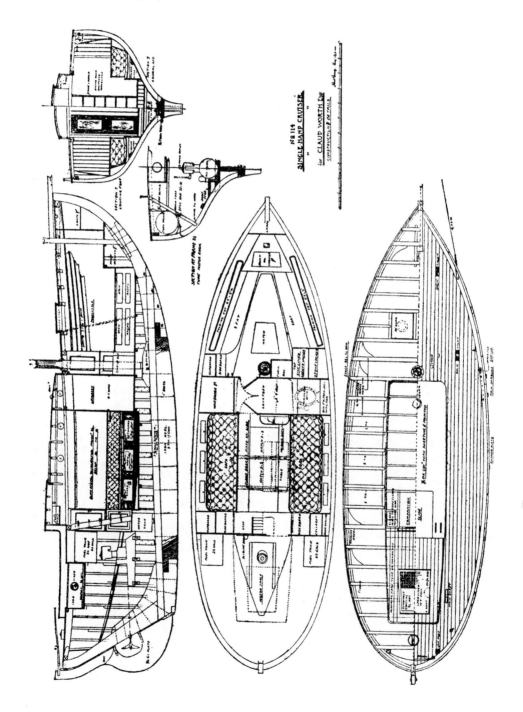

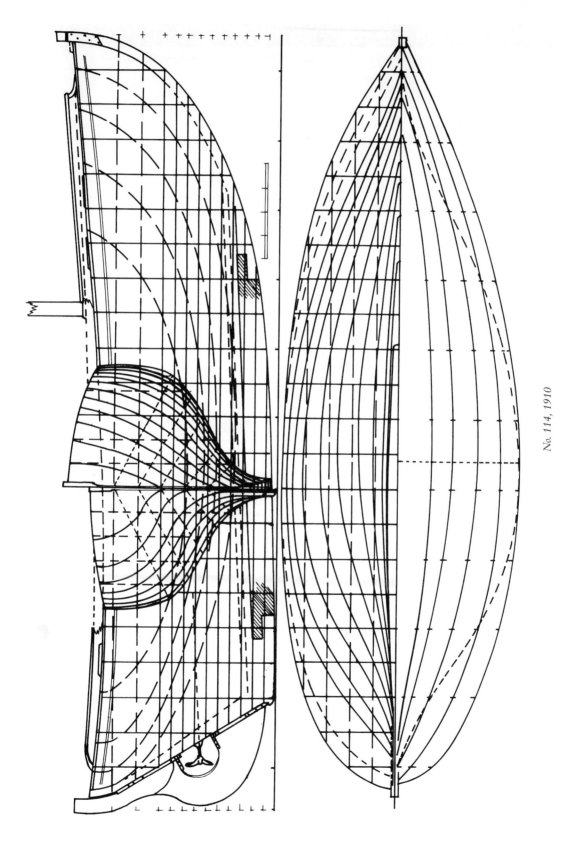

No. 114, 1910

SINGLE-HANDED KETCH 1915   No. 147

LOA: 36ft 7in	LWL: 30ft 0in	Beam: 10ft 2in	Draught: 5ft 0in
Sail Area: 665sq.ft	Disp:11 tons	Ballast Keel: 2.7 tons	

One of only a handful of ketches designed by Strange, no record has been found of this vessel having been built. The remarks which accompanied the publication of the design in *The Yachting Monthly* are illuminating.

Dr Cooke had seen and admired the Norwegian type of cutter which had been designed by Albert Strange for Dr Claud Worth and wished to have a vessel similar in hull, but rigged as a ketch. The change demanded a greater overall length, in order to accommodate the rig and the canoe stern added to the original design has preserved all the seagoing qualities and made a handsomer and easier hull above the water. The forebody remains exactly the same, but the afterbody is a foot longer and the displacement is increased by 18cwt. The sail area is also a little larger, without a topsail, but remains extremely moderate in size for the very powerful hull.

A study of the lines in comparison with those of No. 114 for Claud Worth (page 181) bear this out.[1] The sail area of 665sq. ft is indeed small on a yacht of 11-tons displacement and it is difficult to see how it could be much increased within the confines of a ketch's sail plan. But it conforms to Strange's contention that 'for the ordinary amateur a sail of 300sq. ft area is quite a handful on a small yacht in anything of a sea way'. A view surprisingly prevalent at the time was that the presence of an auxiliary

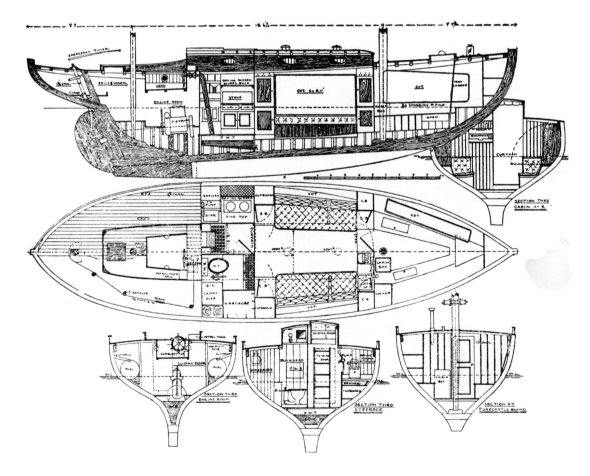

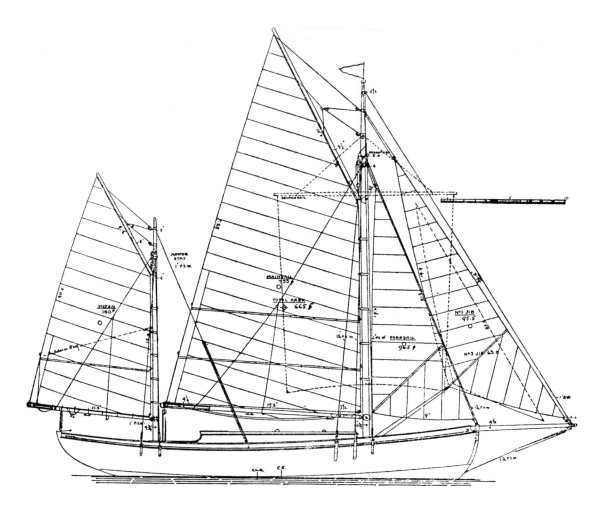

motor mitigated the need for light weather and windward performance. (See page 80, column 1.)

The commentary points out that the wheel can be locked in position and that 'although no trouble is anticipated with the steering gear, an emergency tiller can be fitted. The dinghy (9ft 3in) can be stowed aft when required. The engine is to be a 7-9 Kelvin, starting on petrol, the carburetter being in the cockpit.' The squaresail was favoured by Claud Worth and others at the time for long passages, and has much to recommend it.

The drawing on page 185 is included for interest. At the time the ketch design was published, a correspondence was in progress concerning systems of perspective drawing. A contributor, writing under the pseudonym 'Siberian'[2], had explained a system in some detail which was criticised on various counts, one of which was its complexity. Strange defended it as the best representation of perspective (as opposed to isometric projection, which has no vanishing point) and 'Siberian' sent in this example, for which he happened to use the Single-Handed Ketch, stating that it took him 4¾ hours and was not therefore unduly long-winded.

---

1 For Strange's discussion of the effect of different stern types on the buttock and water lines of the afterbody, as is evident in the comparison of this design with No. 114, see page 33f. and Fig. 6.

2 Believed to be Rear-Admiral Turner, who developed the Metacentric Shelf Theory.

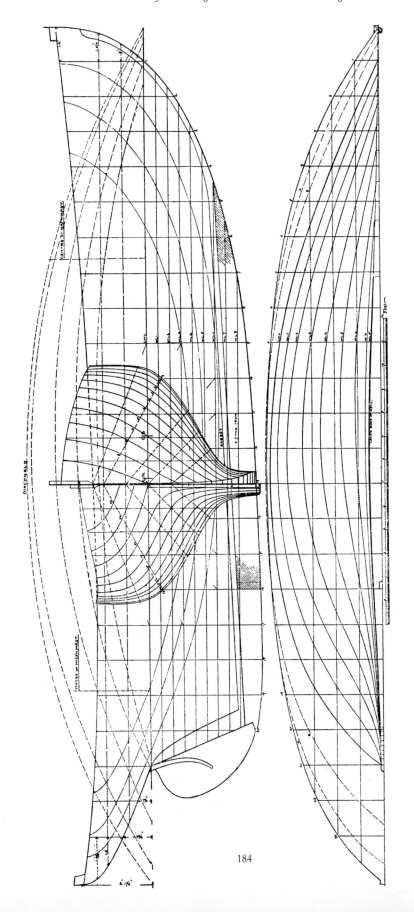

Single-handed auxiliary ketch designed for Mr G H Cooke.

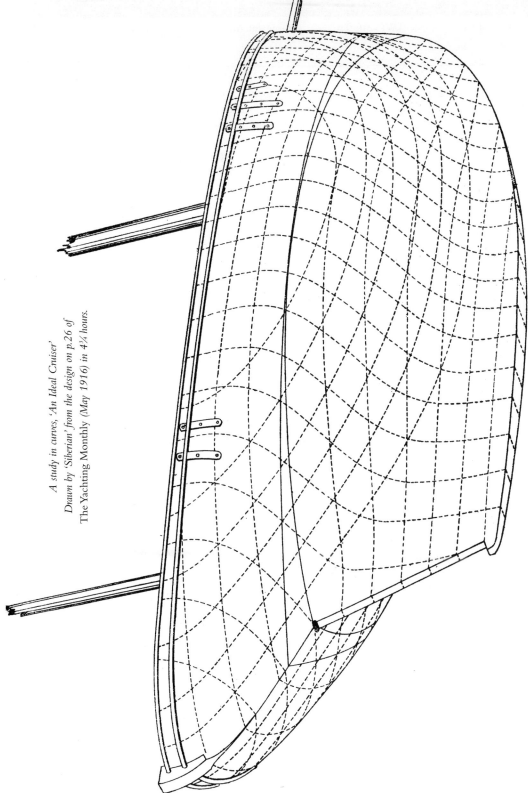

*A study in curves, 'An Ideal Cruiser'
Drawn by 'Siberian' from the design on p.26 of
The Yachting Monthly (May 1916) in 4¾ hours.*

TUI 1915   No. 145

LOA: 30ft 0in	LWL: 24ft 0in	Beam: 8ft 7in	Draught: 4ft 11in
Sail Area: 614sq.ft		Ballast keel: 2.4 tons	

Strange designed *Tui* in 1915 for a Mr. Walter Hopkins. This gentleman had previously owned a number of small craft including an interesting 4-ton sloop, *Tarana*, from the board of Norman Dallimore. There is no evidence that *Tui* was ever built. After the war, Hopkins owned two further boats designed by Dallimore.

The editor's comments which accompanied publication of the *Tui* plans in *The Yachting Monthly* in April 1917, explain the design brief.

Tui is intended for cruising in summer and winter and will have her headquarters on the East Coast. The instructions called for a stiff, seaworthy boat, with good accommodation for three persons, with snug yet fully effective rig, which should be within the power of a single hand and yet give a good turn of speed in moderate breezes. The concentration of sail should ensure this result, and the powerful type of hull section will enable the area to be carried with ease in ordinary cruising weather, and the generous amount of freeboard should give ample buoyancy and dry going in a sea way.

Such a type of cruising boat is very suitable for use in many other districts outside the East Coast and is amply sturdy enough for a cross-channel passage to Allied ports when the 'freedom of the seas' is once more established. There is ample headroom and elbow room in every part of this compact little hull, whose specification shows that the construction is on the strong side and capable of meeting all the emergencies of general cruising, such as taking the ground, lying in dry harbours or alongside fishing craft.

The frames are all steamed, and are 1¾in by 1⅜in, spaced 7in. The outside plank will be ⅞in, the deck 1⅛in in narrow planks, Kauri pine. Deck and bilge stringers, deck beams and carlines are rather above Lloyd's requirements. Galvanised wrought-iron floors connect each alternative frame amidships, while at the ends wood floors receive the upper ends of iron keel bolts. There is 5ft 11in headroom under the small skylight, though the yacht is practically flush-decked. The owner is at present on active service and will doubtless find the change from petrol propulsion to that of the pure if fitful winds of heaven a very pleasant one.

On page 54, and Fig. 25, page 55, Strange explains how the Wave Line Theory first propounded by Scott Russell was modified by Colin Archer. Although Strange may have applied this theory to the curve of areas of other designs, *Tui's* design is one of very few instances which show the circle projecting the curve of versed sines forward and the trochoid aft.

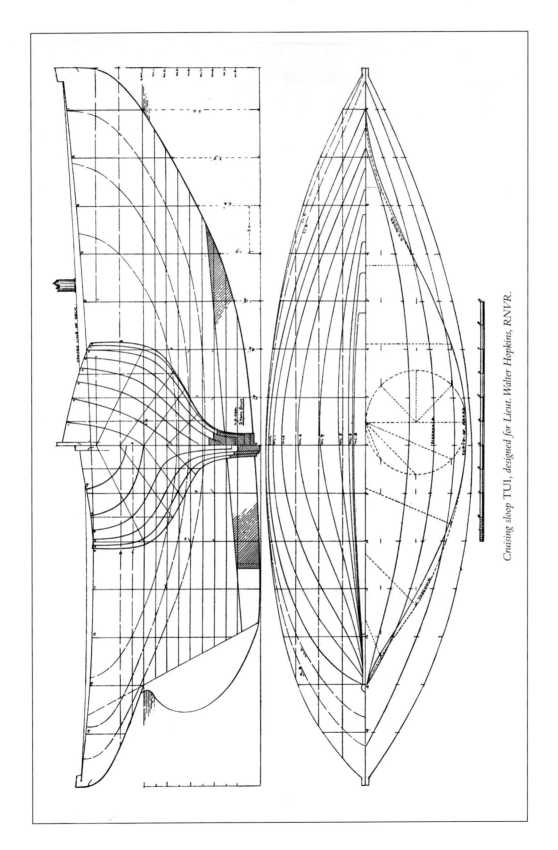

Cruising sloop TUI, designed for Lieut. Walter Hopkins, RNVR.

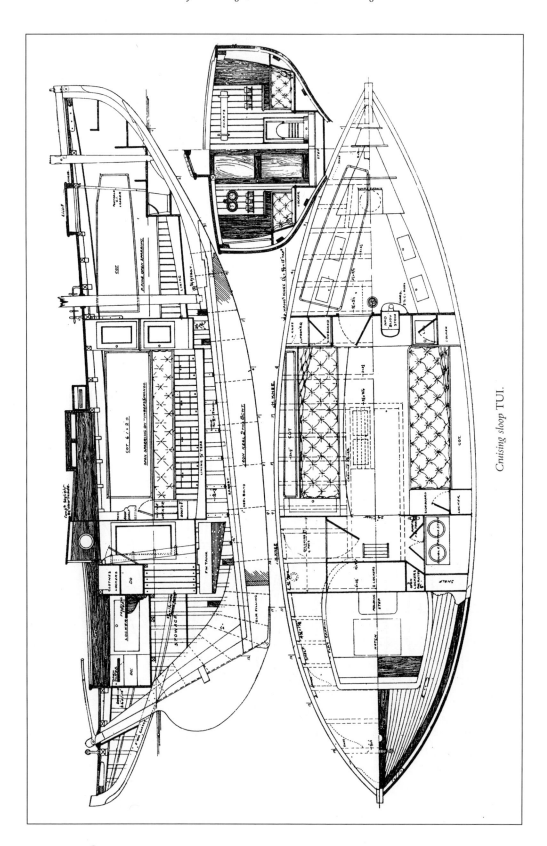

Cruising sloop TUI.

TWO DESIGNS TO THE PROPOSED
BOAT RACING ASSOCIATION RATING RULE 1916

FIN OR CENTREBOARD TYPE	LOA: 22ft 0in	LWL: 20ft 0in	Beam: 6ft 3in
Draught: Bulb fin 4ft 8½in centreboard 5ft 6in	Sail Area: Bulb fin 368.6sq. ft centreboard 370sq. ft	Displacement: Bulb fin 3,460lbs centreboard 3,632lbs	

KEEL TYPE	LOA: 26ft 7in	LWL: 17ft 6in	Beam: 5ft 9in
Draught: 4ft 0in	Sail Area: 380.25sq. ft	Displacement: 4,096lbs	

The Boat Racing Association had been formed in 1912 by a group wanting to encourage small boat racing at reasonable cost and, to this end, invited Major Malden Heckstall-Smith to devise a rule which would allow racing between a variety of types. The B.R.A. had a very short existence. It was merged with the Yacht Racing Association late in 1919. The rule, however, was adopted for the National 18ft and 12ft classes and there is evidence that it was also used in the design of at least one other class, the Salcombe Saints, designed by G L Watson & Co. in 1919.

The two designs included here demonstrate Strange's interest in rating rules and their effect on design – an issue as hotly debated then as now and one which intensified during the First World War, since the energy of those involved was of necessity channelled into theory rather than actual building and racing. They were published in December 1916 as the 14th in a series contributed by professional and amateur designers alike, including Heckstall-Smith, Mylne, Nicholson, Linton-Hope, E P Hart, and Morgan Giles, in which the strengths and weaknesses of the proposed B.R.A. rule were discussed at great length.

Strange's remarks accompanying the designs, explain the exercise. While approving the rule because it would be 'more likely to produce a seaworthy type and a drier boat', his reservation was that...

> *It will be found... likely to produce a more fixed type of yacht and so far is a 'restriction on design', thus placing this rule with nearly all others as a type producer.*
>
> *The design of normal type [see page 191] is intended for light and moderate winds, without having too much initial stiffness. The 'boat' type [see page 190] is intended... to test the difference between a bulb fin and a centreboard boat.*

His comments on the bulb-fin keel strike a familiar note across 80 years of development in hull design:

> *A careful comparison of the three types clearly shows what a tremendous advantage in sail-carrying power the bulb fin possesses and it seems almost certain that where conditions allow, this type must prevail over all other types, from extreme light displacement to comparatively heavy weight.*
>
> *I am sure that this is a point that wants watching and checking by some restriction, if the general dislike of fin bulb boats is to be taken into consideration. This dislike is, I fancy, very real and widespread.*
>
> *There seems to be no necessity to send sail plans with these designs. It is evident that the Marconi-Bermuda has come to stay so far as racing boats are concerned and these boats, if built, would of necessity have this rig.*

Strange had contributed several more letters to the debate by the time of his death six months later, and it continued unabated into 1919. It should be said that his are some of the lighter-hearted and more digestible contributions to a correspondence which at times became abstruse and very mathematical. He never lost sight of the desired result, which was to him a moderate, popular, and economical boat. In a letter of May 1917, having suggested a much simplified rule, he launches an attack on the increasingly complicated mathematical formulae being propounded by others:

> *Why such a sensible rule has never been adopted is quite beyond my feeble power of comprehension. I can only imagine that the persons who gather themselves together to concoct rating rules are merely mathematicians, and we all know that mathematicians, like politicians, will never willingly countenance any methods or statements that can be clearly grasped by ordinary ignorant or simple people at first sight.*

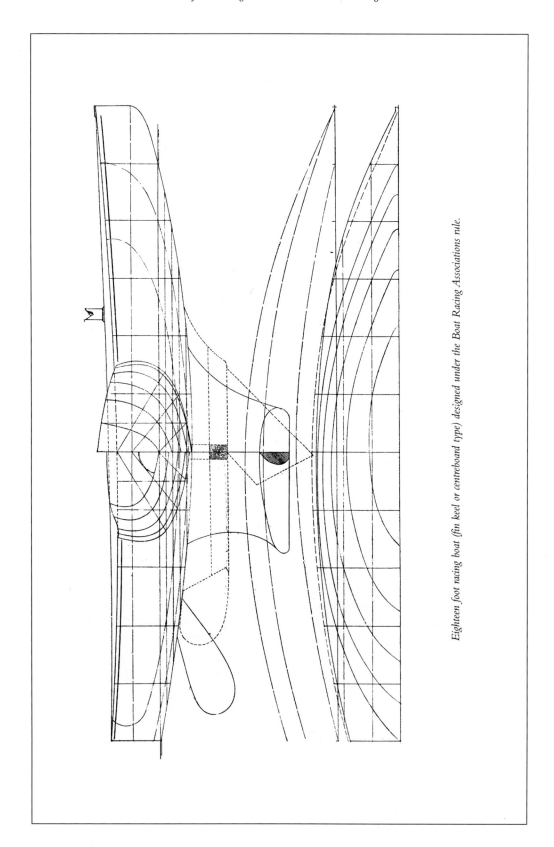

Eighteen foot racing boat (fin keel or centreboard type) designed under the Boat Racing Associations rule.

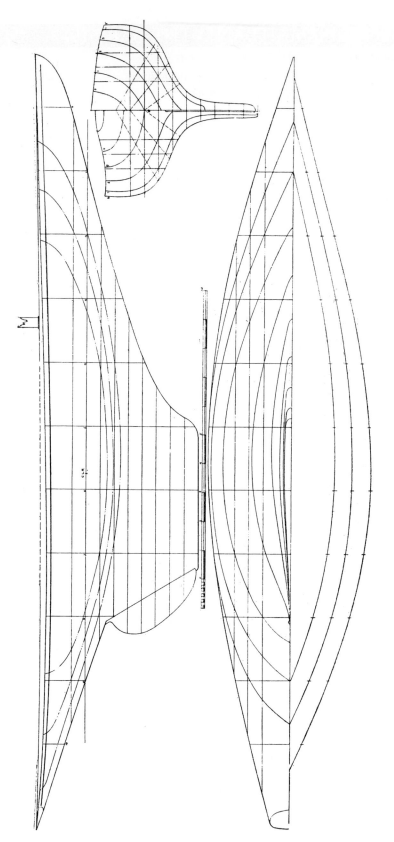

Eighteen foot racing boat (keel type) designed under the Boat Racing Associations rule.

THE NURSEMAID AND THE BABY 1917

NURSEMAID

LOA: 28ft 0in	LWL: 21ft 0in	Beam: 7ft 6in	Draught: 4ft 2in
Sail Area: 505sq.ft	Displacement: 9,184lbs		

BABY

LOA: 28ft 0in	LWL: 21ft 0in	Beam: 7ft 8in	Draught: 4ft 5in
Sail Area: 430sq.ft	Displacement: 7,820lbs		

While the B.R.A. debate was still in full swing in the pages of *The Yachting Monthly*, its editor held forth at some length, in March 1917, to propose the idea of a 'National Nursery Class'. He may well have been anxious to draw his contributors away from the endless discussion of a rating formula, but he also saw a genuine and urgent need to establish a sound basis upon which the sport could grow rapidly among the young and the less well-off, once the War was over.

To this end, he argued that a cruising boat which could provide good racing in class was needed. Since a rating rule inevitably produced boats in which everything was sacrificed to speed, this proposed class was to be restricted. The type of restrictions envisaged is demonstrated by the two designs submitted to the magazine by Strange and published in the June 1917 edition.

The full text of Strange's commentary follows, since it includes, as usual, interesting observations on life in general as well as the matter in hand; a brief view on the effect of the various rating rules since the introduction of the Length and Sail Area Rule of 1888; and his opinion of what was a good boat for 'the young men returning from the War'.

In fact, the Nursery Class idea never took root and one-design classes flourished; the restricted-class concept remained more attractive to the designer than to the man of limited means wanting a boat to cruise and occasionally race.

As one critic of the idea wrote...

The idea of small racer cruiser classes such as the late Albert Strange advocated, is one with which I am in sympathy altogether; but I have very faint hope of seeing such a scheme take hold, to any great extent, among the youth of this realm. It seems to me that small sailing cruisers, so desirable for the young, are generally advocated by the old or middle-aged who remember how small cruisers flourished in the days of our youth and overlook the advance of ideas in yachting, as in everything else, whereby the present generation is leaving us behind ... I do not see youth take to the sea for sport or pleasure at all generally, either for racing or cruising in these days of motorbicycles and cars.

The "Nursery" Class: an Effort to Assist

BY

ALBERT STRANGE

WE are often very much indebted to M.I.N.A[1] for helpful suggestions, not only for our sport, but occasionally for our moral good. He has seldom touched upon a subject so fruitful in every sense as the Nursery Class—a class, I take it, that will combine in one boat the possibilities of racing and cruising. These possibilities were easily recognizable in the days before the advent of the length and sail area rule, and still continued to be practised for a good many years after the sweeping victory of the light displacement craft developed under that rule—in canoes and canoe yawls and very occasionally in one-design or restricted classes.

But it is a sad though actual fact that for the last fifteen years cruising and racing have become absolutely divorced in all classes under 12-metre size. The racers were, so it is alleged, not fit to cruise in, and the cruisers could not be raced, even when they were ex-racers, with the least hope of success against newer craft, or against each other. No method, save the casual and haphazard one of handicapping, has ever been devised that could bring various sizes and types of yacht together on a common or just footing—of even relative equality— nor does one seem likely to be invented, though many praiseworthy attempts are still being made.

The result was, until war took our young men away from their yachts, an absolute divorce between racing men and cruising men, both in habits and outlook. The cruising men increased in numbers, and, as I know well, enlarged their ideas very considerably as to the shape, accommodation, sail plan and general effectiveness of their yachts. The racing men, though far fewer in number, moved too, but were bound up in the fetters of the International Rule, and, in spite of the noble attempt made by Mr. Rathbone and Mr. Linton Hope to initiate a cruising class of 6-metre boats, cruising fell off almost altogether in class boats. I can only at the moment recall Mr. " Jack " Little cruising and racing in Nargie—a solitary instance of cruising in a racing 12-metre yacht. There may have been others, but I do not think there were many. Something had come between the two different sets of men, one can even now note a scarcely veiled contempt for the cruiser in the writings of the racing man—which attitude is returned with full interest when the cruiser refers to the racing man or his yacht, either in conversation or correspondence. This feeling ought not to exist among yachtsmen, and there was certainly less evidence of it thirty years ago than there was in 1914. There can be little doubt that the development of the racer, since the L. and S.A. rule came in, tended to produce this rift, and it also enlarged our knowledge of the retarding effect of weight on speed. Then came the well-intentioned International Rule, giving back its chance to weight, but so cramping both beam and draught that we now have, in the smaller classes, a type less effective all round than even the old " plank on edge " boats were.

Many have felt and argued that a rule which produces a good type on the

1 MINA is the pseudonym under which Herbert Reiach, the editor of *The Yachting Monthly*, wrote.

whole in the large classes has a contrary effect in the small ones. The truth of this contention is now more fully realized. I think it cannot be doubted that there is a point—which seems to me to make the most extreme experiments. These chances are less obvious and less easy to take in the design of larger yachts, where natural restrictions, such as depth of available water, difficulties of

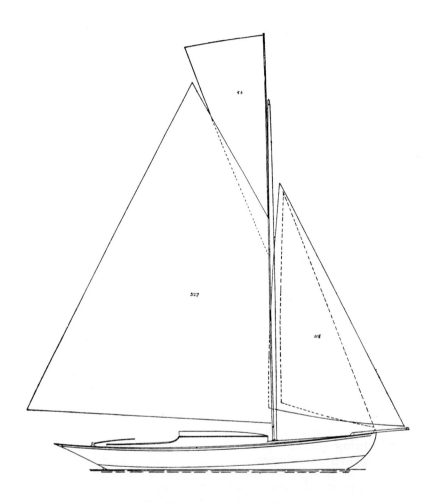

SAIL PLAN OF THE NURSEMAID

Scale : ⅛ in. = 1 ft.

Total Sail Area, 505 sq. ft.

lie between the ten-metre and the seven-metre boat—when restriction on beam tells against the production of a good all-round wholesome boat, and leads to an undesirable type. But there must be restrictions—especially in the smaller sizes, where there is every chance to rig, and the question of expense, all exercise a restraining effect on extravagant adventures.

If restrictions have to be submitted to we ought, as sensible men, to see that they tend in the direction of practical usefulness, while at the same time they should not prohibit any desirable freedom of design. A " one class one design " in which nothing is left but a cast-iron form of hull and sail plan, can have but small attraction for the average man, not to speak of the yacht designer, who desires to live and to create —just as any other artist, poet or painter, desires to express himself. Only the yacht designer—the racing yacht designer especially—is not given many chances. A very few mistakes seal his doom, while a painter can go on producing unpopular pictures with far less danger of imminent and instant extinction. Some people like queer pictures, but few people tolerate queer boats unless they win races; but in succeeding in this they fulfil the object of their creation.

I suppose there will always be racing men pure and simple, and cruising men who would never race, and hate the idea. But between these two sorts of individuality there are very many yachtsmen who love to cruise in a smart boat, and who would be eager to race on summer evenings or week-ends if they could race on their own cruisers on fair terms, or terms which were not humiliating, as most racing between the out-and-out racer and the out-and-out cruiser always is and always must be.

So we must accept restrictions, and as every rule yet devised is full of restrictions—some more full than others, but all have them—(a rule is in itself a restriction)—let us consider whether restrictions can be usefully employed in producing a class of boat that will be comparatively inexpensive as compared with a racer of the same size, and, what is more important, a class that will encourage and foster the sort of sailor-yachtsman we all have a vision of in our minds—adventurous, hardy, self-reliant, and resourceful. All these qualities exist in the average man. War has proved this conclusively, if it needed to be proved, and yachting is, of all sports, the one to develop and perpetuate them when war has killed itself.

Racing develops valuable qualities, and therefore must be preserved. But racing has undoubtedly militated against the fullest development of the full sum of sea virtues, especially since professionalism has grown to be the necessity it is in match sailing in larger craft. The nature of small class racing makes a demand on the qualities of quickness and perception, but it hardly touches endurance, because a few hours cover the course. If the Nursery Class is to fulfil the objects we all desire, racing must continue, but the races must be longer. They must be from port to port, and the results must depend upon skill in navigation as much as upon skill in getting the best out of a boat, out of the tides, and the seizing of an offered opportunity. A race from Erith to Burnham, from Burnham to Harwich, from Harwich to Southwold or Lowestoft is the sort of thing that will answer the purposes we are proposing. The distances, though not great, would give long trials of seamanship. Such races would really be cruiser races, and afford excellent drill in making and shortening sail, navigation, anchoring, night sailing, harbour taking and clearing, rule of the road, and the hundred opportunities of learning the true inwardness of sea sailing.

This kind of racing demands a special class of boat, one possessing a cabin, however small the boat may be, and the two sketches accompanying this article give an idea of what seems to me to be a suitable type for the work. They show no new points, I know, but the cost of a yacht of this kind could be borne without hardship by a partnership of two or three men, who would find just enough room (but no more) in which to spend a few nights at sea, or a few weeks' summer holiday cruise. They could also be raced at week-ends over short courses, and, though not thoroughbreds, have enough sail to ensure sport in any wind above a light air. I believe that the Nursery Class *must* be of a cruising type, *must* be as small as possible, and *must* be devoid of costly features if it is to produce the results we desire.

These are only two out of an immense variety that could be produced under the proposed restrictions which will follow, and though they are small for cruisers, they are a good deal larger in dimensions, displacement, sail area, draught, and cabin room than Sheila,[1] the accounts of whose cruises in the Irish Sea and on the West Coast of Scotland so much delighted us in the earlier years of the YACHTING MONTHLY. They are quite the western end of the English Channel, but for almost every other station in Great Britain they would serve well. I do not think that they are perfect, and they are not put forth with any other

SAIL PLAN OF THE BABY

Scale : ⅛ in. = 1 ft.

Total Sail Area, 430 sq. ft.

big enough for young men, yet not so big as to be expensive. They are saleable when their purpose is accomplished, and the class is capable of use in most districts. A yacht of deeper draught in proportion to her length would perhaps be more suitable for Scottish waters and idea than that of eliciting criticism and probably better examples from other designers.

Of the two I think The Baby would be the faster in fresh winds and smooth water, but the larger sail area of The Nursemaid would tell in light winds,

1 See page 10.

while in a bit of a jump, so often found with a whole-sail breeze at sea, she should, by her greater weight, take the lead. I am not sure, of course, but I do not think that very much greater sail area per 100 lbs. would tend to produce a better boat, but here is a " variable " that is most easily adjusted without in-

justice and without necessarily causing a large number of boats to be scrapped in consequence of *hull* alterations. The cabin requirements, the relation of free-board to draught, and the minimum area of lateral plane secure at least a habit-able boat. The proposed restrictions are appended :

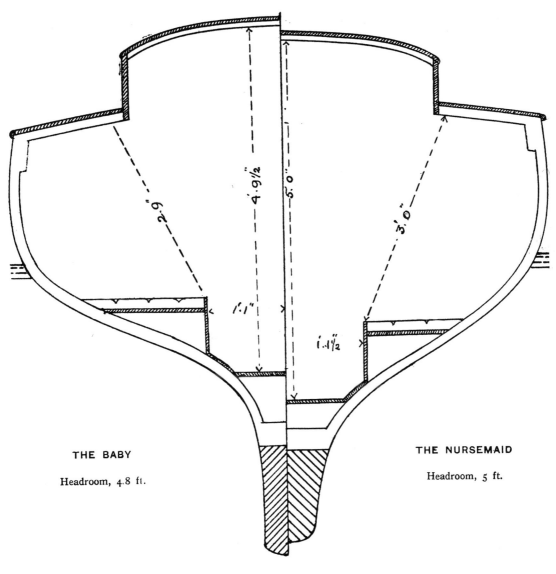

THE BABY

Headroom, 4.8 ft.

THE NURSEMAID

Headroom, 5 ft.

CABIN SECTIONS

Scale : 1 in. = 1 ft.

PROPOSED RESTRICTIONS FOR A NATIONAL NURSERY CLASS

HULL RESTRICTIONS.

Length.—Not to exceed 28 ft. over all. L.W.L. not to exceed 21 ft.

Draught.—Not to exceed twice mean freeboard measured at fore and aft end of W.L. and at lowest point.

Area of Lateral Plane.—Not to be less than 45 sq. ft. Rudder to be hung on sternpost at aft end of W.L. Front edge to be at least 80 per cent. of total draught in length.

Beam.—Not to be less than 7 feet.

Ballast.—Thirty-five per cent. of the total displacement to be in the form of an iron keel. Inside ballast, if required, may be of lead, securely stowed.

Sail Area.—5.5 square feet of fore and aft canvas to be taken for each 100 lbs. total displacement. This area must be in three pieces, one of which must not be less than 10 per cent. of the whole amount. A spinnaker containing 20 per cent. of the total fore and aft area is allowed in addition. Any rig may be adopted that has three sails.

Square stern boats not to exceed 21.6 W.L. and 24 ft. O.A. All other restrictions to apply as in boats of 28 ft. over all length.

CABIN RESTRICTIONS.

Least Headroom.—4 ft. 9 in. Least width between seats 24 in. Flat of cabin sole not to be less than 18 in. in centre and 12 in. at each end of a length of 6 ft. Cabin top not less than 6 ft. 6 in. in length.

Sleeping Accommodation.—At least two berths or folding cots; 24 in. wide in centre by 20 in. at either end, with 2 in. mattresses, and blankets.

One compass, one stove, ½ gal. oil, 4 galls. water, one 30 lb. anchor, 35 fathoms of 2½ in. rope cable, one pair of oars, one riding light, one lead line (4 lbs)., to be carried in all races. In long distance races a lifebuoy or belt for each of the crew, and a folding or other dinghy to be carried.

SCANTLINGS.

Stem & stern posts, 3¼ in. sided; coamings, ¾ in. throughout; cabin roof, ⅝ in. covered; main keel, 3½ in. siding throughout; planking, ¾ in. finished; timbers, 1½ in.×1 in., spaced 6 in. centres; shelf, 6 sq. in. sectional area; bilge stringer, 4½ sq. in. sectional area; deck, ⅝ in. thick, covered with canvas; Beams: three main beams, 2¾ in.×2 in. at centres; other beams, 2½ in.×1¾ in. at centres; half-beams, 1¼ in.×1¼ in. Spacing of beams, 12 in. centres; five wood floors; 3 in.×5 in. throats; galvanized iron floors on alternate frames, 1½ in.×½ in. throats; 16 in. arms, tapered to ¼ in. at points.

Doubtless many other suggestions can be built on these, and it will be very interesting to see attempts made to " break through " or evade the conditions. But I hope that competent critics will come forward with amendments and criticisms, so that out of it all may come something to fill the want required so urgently in the near future.

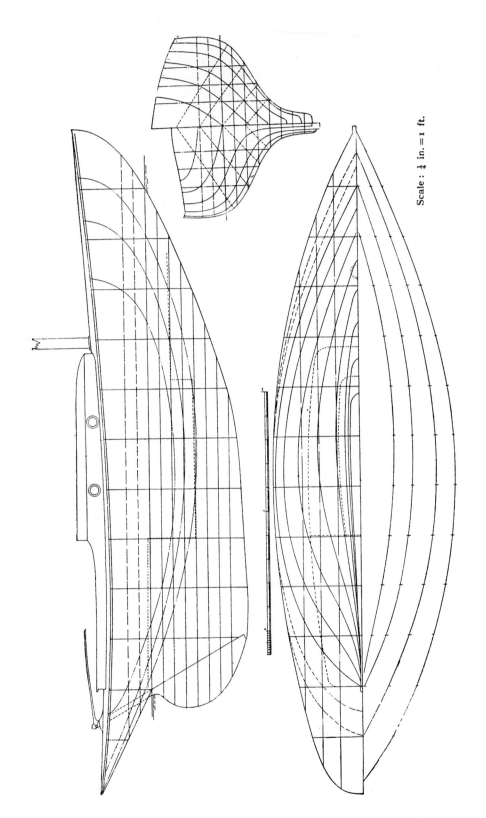

THE NURSEMAID

Scale : ¼ in. = 1 ft.

L.O.A., 28 ft. ; L.W.L., 21 ft. ; Beam, 7.5 ft. ; Draught, 4.2 ft. ; Displacement, 9,184 lbs. ; Area L.P., 62.2 sq. ft.

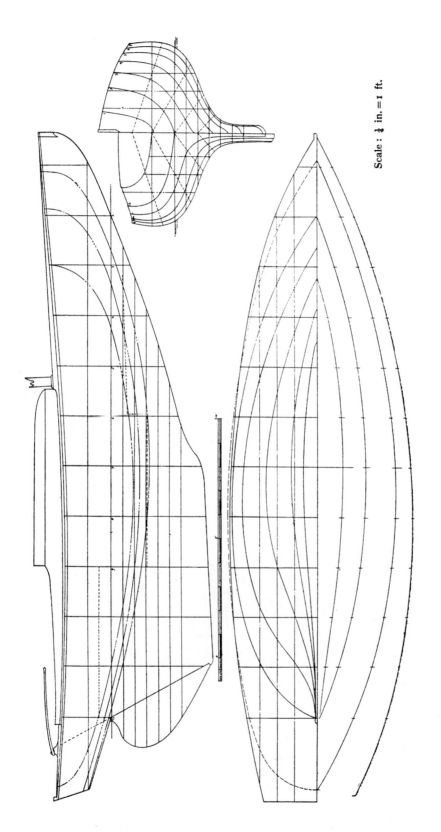

Scale: $\frac{1}{4}$ in. = 1 ft.

THE BABY

L.O.A., 28 ft. ; L.W.L., 21 ft. ; Beam, 7.7 ft. ; Draught, 4.4 ft. ; Displacement, 7,820 lbs. ; Area L.P., 65.4 sq. ft.

THE JILT 1917   No. 150

LOA: 30ft 0in	LWL: 26ft 0in	Beam: 8ft 5in	Draught: 4ft 9in
Sail Area: 530sq.ft		Ballast keel: 2.38 tons	

This short note accompanied publication of the plans of *The Jilt* in *The Yachting Monthly* in December 1918.

THE JILT: A LATE STRANGE

Sir, – I send you the plans of a boat designed for me by the late Albert Strange. I should like to hear what you and your readers think of her. I intended her for short-handed use on the South [Coast] of Ireland.

G C

Strange produced a particularly elaborate set of drawings for Captain G Clarke, running to ten sheets and including a table of offsets. He completed the work in Febraury 1917. At that time, the prospective owner was based in Cork, Ireland and owned *Psyche*, a shallow-draught Humber Yawl-type cruiser of six tons, also designed by Strange. (See p. 122 and JL p.115.)

Later, he moved to Liverpool and must have decided not to build, because he bought another shoal-draught boat from the board of M Treleaven Reade, a talented local amateur. This change of plan explains the use of *intended* in his enquiry to *The Yachting Monthly*.

It is disappointing that there were no replies to Clarke's query. This was unusual, as there always seemed to be someone ready to criticise any design in those days, no matter whose work it was.

There is no evidence that anyone else ever built to this design, which is also a pity, because she would have made a fine and handsome cruising boat.

SAIL AREAS

mainsail	*375sq.ft*
staysail	*84sq.ft*
jib	*71sq.ft*
Total	*530sq.ft*
(squaresail 200sq.ft)	

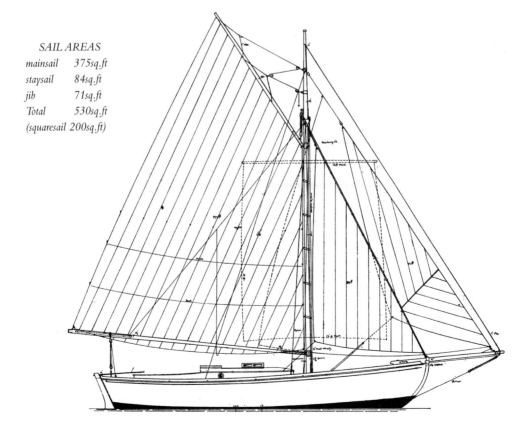

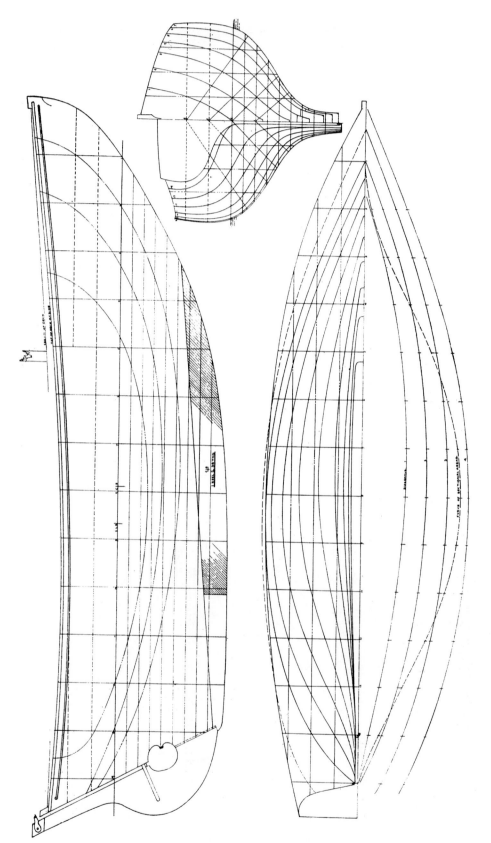

THE JILT, *an auxiliary cutter designed for Captain G A E Clarke.*

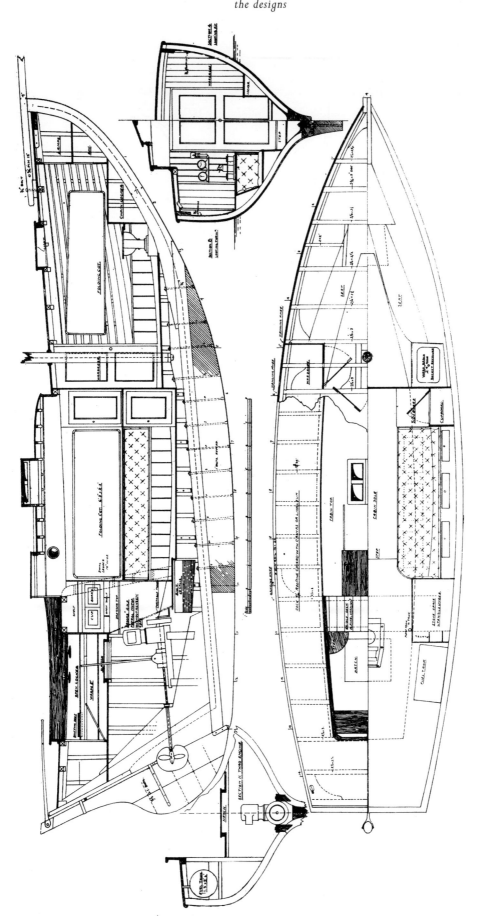

THE JILT, *auxiliary cutter.*

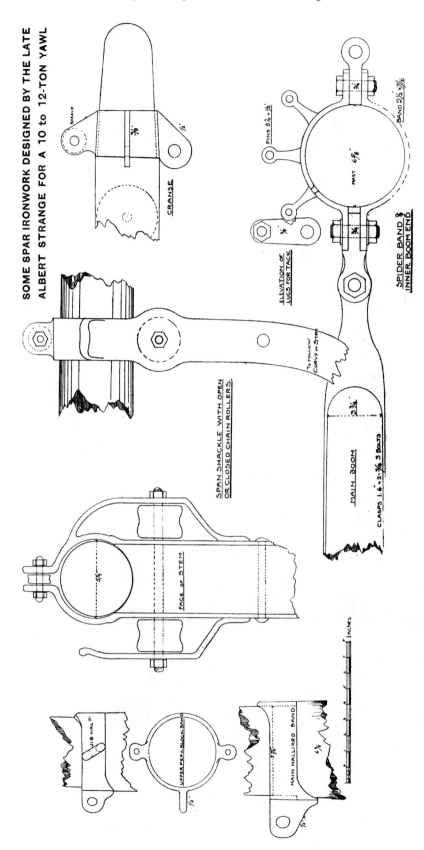

CHAPTER 4

An Old Time Cruiser

The Yachting Monthly, April 1916

Geo. Holmes   *YM April 1911*

The two decades leading up to the First World War had seen a great increase in popularity and rapid development in both racing and cruising. The War years saw few opportunities for sailing, and were thus a time to 'take stock' of previous development and to contemplate future directions.

Strange energetically contributed to this debate and it is fascinating to be reminded of the great strides that had been made in yacht design by the time the First World War forced a temporary halt. It is easy, 80 years on, to look through the designs in this volume and perceive them merely as 'traditional', rather quaint, or lacking in the accommodation we have come to expect from boats, and so on.

Yet, here we have Strange looking back 60 years and regarding a yacht of the 1850s with a stronger sense of its archaism than we have for the designs of the early twentieth century. For Strange, the greatest advance in the technology of construction was the introduction (and acceptance) into general use of the external ballast keel. For us the advances are so dramatic, one would expect the yachts of Strange's day to appear as veritable dinosaurs. But they do not because, in terms of actual, practical, everyday cruising use, advances in construction and materials have not made such a dramatic difference as one might have expected. At the same time, advances in design form have not all produced unqualified advantages for the cruising yachtsman, and thus, the level of refinement in cruising-yacht form reached by, say, 1920, still has much to recommend it.

Strange's article provides a fascinating perspective on mid-nineteenth-century yacht design and, as always, on the quality of life and social attitudes. His unfailing interest in humanity and the world around him shine through his words.

Although not a log of one of his own cruises, it is in 'An Old-Time Cruiser' that Strange expresses so succinctly his view of what a log should *not* be – 'a dull record of times of arrival and departure, with nothing in the way of adventure or observation of human nature set down'. When one reads the two accounts of his own cruises which follow this article, as well as the other delightful accounts reprinted in John Leather's book, it is abundantly clear that he practised what he preached.

YM April 1911

YM April 1909

1 ie 1853.
2 Royal Harwich Yacht Club.
3 Honourable East India Company's Service.
4 Firth passage boat and Firth of Dornoch.
5 Royal Mersey Yacht Club.
6 Vanderdecken was, in fact, the pseudonym of William Cooper, so it would seem that Strange had indeed committed a 'rank injustice'. Whatever the merits of his fiction, Cooper is better known and respected for his manual *Yachts and Yachting*, 'a treatise on building, sparring, canvassing, sailing and the general management of yachts', which was published in 1873 by Hunt & Co.
7 *Idas* was built in 1850, a sturdy little cutter, 22ft 6in long, 7ft beam, possibly clinker built, and fully rigged with topmast, running bowsprit, and square-headed topsail. (These details are taken from *Traditions and Memories of American Yachting* by W P Stephens, page 299).
8 Prince of Wales Yacht Club.
9 Old Measurement. Before the introduction of Thames Measurement in 1855, racing yachts were measured by a 'Tonnage Rule' known as 'Builders Old Measurement'. This was calculated by the formula:

$$\frac{\text{Length along keel x beam x depth of hold}}{96}$$

An Old-Time Cruiser

BY

ALBERT STRANGE

SIXTY-THREE years ago,[1] when England, by reason of her discoveries and inventions, was entering on a period of commercial prosperity that was to place her in the forefront of the nations of the world, there appeared the first number of " Hunt's Yachting Magazine," edited by " Willam Knight, Esq., R.H.Y.C.,[2]" the first attempt to preserve in an enduring form records of the racing, cruising, and construction of yachts, which has culminated in THE YACHTING MONTHLY and in every other periodical, British or foreign, devoted to yachting.

The first number contains sixty pages of letterpress, treating of such subjects as " Our Royal Yacht Clubs," " The Iron Model Yacht Problem "—a fearsome square-rigged craft of triangular sheer plan, warranted to sail ahead or astern without going about; which, according to her inventor, Henry Dempster, H.E.I.C.S.,[3] would infallibly " fetch to windward of the place she looks at from the starting point." There is a long explanation of the manner in which this feat would be accomplished " from the hydraulic action of the fluid upon the peculiar formation of her hull," etc., etc. Then follows a description of " A Visit to Farley's." Farley's was a model yacht emporium where you could purchase built sailing models at prices ranging from £2 10s. to £20, and models of the celebrated America on a scale of a quarter of an inch to the foot, " and also on three-eighths." Then comes " Our Editor's Box " for correspondence. "Notes from the Nore," " Reports of Sailing Matches and Yacht Races," concluding with a Chronicle of the Month, which was only an almanack recording such historical events as one usually finds in these publications, but of a somewhat more nautical character, such as : " August 13, Loss of the Frith passage boat in the Frith of [4] Dornoch and 40 persons drowned, 1809 "; and " August 24, Royal Belgian Yacht Club Regatta, 1852. Commodore Littledale, R.M.Y.C.,[5] saved 32 lives from a ship on fire in Abergele Bay, 1848." There are also, right in the middle of the number, eleven closely printed pages of a serial story entitled " The Channel Cruisers," written by " Vanderdecken," whom I strongly suspect to be William Knight, Esq.[6] himself, though this may be rank injustice on my part, for " The Channel Cruisers " is a hodge-podge of the most melodramatic and unreal nature, plentifully sprinkled with such phrases as " Avast there, villain," " Ah, ah, foiled once more," etc. I have never been able to read it, though it wanders haltingly and with huge monthly gaps through the magazine for four solid years, and then leaves off abruptly unfinished; perhaps to reappear again in after volumes. It is the sort of story one could not read even on a desert island and if deprived of any other literature, but it is not the only specimen of its kind, alas !

The first issue contains a few illustrations—and is a typical first number—the price being one shilling.

Notwithstanding the incubus of " The Channel Cruisers," the magazine found its public, and the cruising man, ever the mainstay of undertakings of this nature, began to contribute. In October appeared an account of a " Cruise from Blackwall to Boulogne and back, in the Idas,[7]" by Vice-Commodore Knibbs, P.W.Y.C.[8] Idas was only 5½ tons O.M.[9]—quite a small craft, though no particulars are given, and the log is a dull record of times of arrival and departure , with nothing in the way of adventure or observation of human nature set

For footnotes, see page 208.

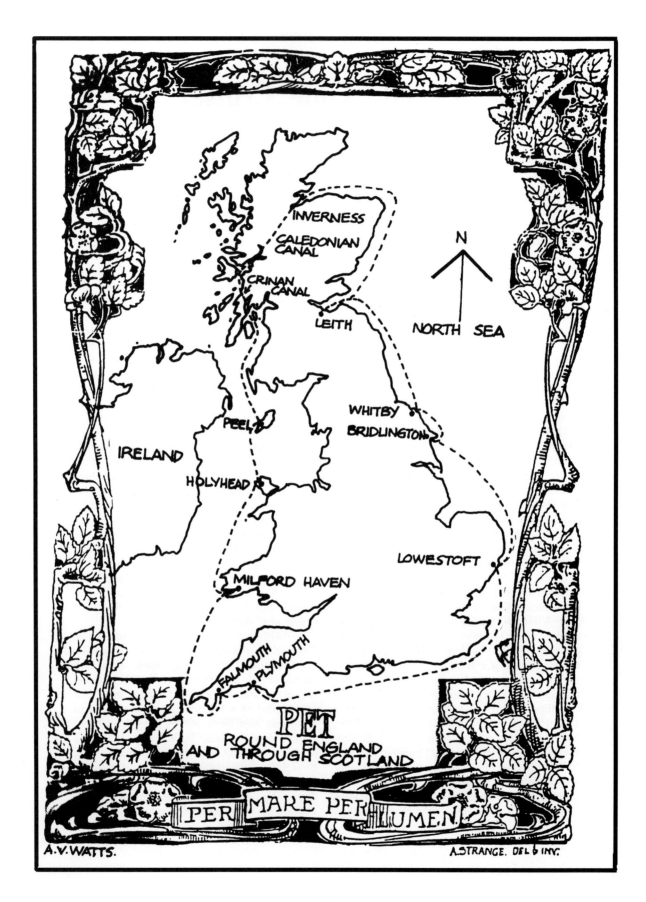

INVERNESS

CALEDONIAN
CANAL

CRINAN
CANAL

LEITH

N

NORTH SEA

WHITBY
BRIDLINGTON

IRELAND

PEEL

HOLYHEAD

LOWESTOFT

MILFORD HAVEN

FALMOUTH
PLYMOUTH

PET
ROUND ENGLAND
AND THROUGH SCOTLAND

PER MARE PER LUMEN

A.V.WATTS.

A.STRANGE. DEL & INV.

down. But it was evidently regarded as a great performance by the Editor, who adds at the end : " Cruising, even to Boulogne in a small craft like the Idas, will not exactly suit gentlemen much given to kid gloves and patent leather boots, but the practical knowledge that may be picked up in a five or six-tonner is by none of us to be despised." I am sure these flattering remarks must have been appreciated by Vice-Commodore Knibbs. This is also the first indication I have come across of the contempt supposed to be felt by the hardy cruiser for any of the refinements of such civilization as existed in the 'fifties.

In January, 1853, appeared the first chapters of " Circumnavigation, or, the Log of the Pet," by R.E.H. Written in clear vigorous English by one who was evidently a " scholar and a gentleman," this log exactly sets the style in which such reminiscences should be written. The only possible fault that can be found with it is for the very meagre amount of actual detail about the yacht and its fittings, though I suppose all eight-ton cutters of that day were practically exactly alike—oversparred, overcanvassed and overballasted, in imitation of the contemporary racing yacht. What few details are granted are offered in a very apologetic way, with the excuse that they are inflicted because a " sailor loves his craft." These are the author's words :

" The Pet is a Poole boat, built by Wanhill, very deep, drawing six feet and a half, very sharp, and very much oversparred. Her boom projects seven feet over her taffrail, and her other sticks are in proportion, or rather in similar disproportion. She carries seven tons of iron ballast, and in her cabin you sit, move, and live a long way under water. She has shown herself a good little sea boat in many a rough berth; in forereaching we did not think much of her, but in working to windward we seldom met anything that could take the wind from her. She is painted black, and coppered, and has a remarkably gentlemanlike and shipshape appearance, especially when with her balloon jib and enormous main topsail set she swaggers away to leeward, her sails looking like a great white albatross and her little black hull like a mackerel in its talons."

This is the brief array of facts, and when I began to figure out the kind of eight-

tonner she would be, it was necessary to go further afield to get any sure foundation on which to work out the design here given. Her other dimensions I excavated from letters in later volumes of " Hunt's " —some correspondents praising, others blaming, the reckless Divine, for it turned out that the initials were those of the Rev. Robert Edgar Hughes, M.A., Fellow of Magdalen College, Cambridge, though as far as I am able to glean, history is silent as to any other facts concerning him. Later cruises show that his brother, J. W. Hughes, was an officer in the " 66th Regiment," and that R.E.H. was a member of the Royal Thames Yacht Club.

As I fancy I have remarked on another occasion, boats are surprising things, and the design shows with a fair degree of accuracy what an eight-tonner was like in the hull in the year 1850. So great is the force of habit that it is possible I may have given her more freeboard than was customary in those days, when a 25-tonner had only 3 ft. 9 in. forward and 2 ft. 6 in. aft. The amount of iron ballast her owner affirms she had—seven tons—seems impossible, but this weight could be carried in a yacht of this shape, though it must have taken up a lot of room, being all inside. Ballast on the keel was then regarded as a direct invitation to disaster, and was shunned by cruisers, though racers (the Idas was a racer) might have had three or four hundredweight outside, but no more. The sections, the sheer line, and the shape of the immersed longitudinal section are accurate representations of the Wanhill yachts, and though clogged up with an unnecessary area of wetted surface, which accounts for lack of speed in reaching, the general shape enables us to understand how easy and steady these yachts were when hove to. In fact, the aim of the yacht-builder in those days (there were few or no yacht designers) seems to have been directed to this one point. The yachts were terribly slow in getting under way, and would hang for several minutes, or even have to be given a stern board, before paying off.

The sail area must have reached from 1,100 to 1,200 square feet all told. Needless to say, a trysail was an invariable item of their equipment, and was constantly used. The hollows in the buttock lines at the counter were common, even in Fife's boats of that date, and are partly due to fine waterlines aft. The midship section was always in the centre of the yacht, and the centre of buoyancy either in the centre or slightly ahead of it. There was good reason for this, though it was not adopted on any scientific basis. But it is a fact that in the very latest examples of the product of the very latest of our racing rules, that is to say, in the six metre boats, the position of the C.B. gradually moved itself ahead with advantage as the displacements grew larger, though, of course, not so far ahead as in the Pet. And odd as the yacht seems when viewed by modern eyes, we must not forget that the Neptune[10] is but a development to its logical conclusion of the type exemplified in the Pet, and is not a different type. Her owner considered her a "gentlemanly looking" vessel. (I wonder, could he return from those cold shades where we are told there is no more sea, what would be his description of the appearance of a canoe-sterned cruiser.) The crew of the Pet consisted of four persons, Mr. Hughes and his brother aft, and two paid hands forward, "lads" Hughes calls them, one rated as boatswain and the other was cook. This of course will be considered a hefty crew for an eight-tonner, but think of the spars and sails, and remember the fact that the yacht was often kept at it day and night.

Leaving Lowestoft on the afternoon of July the 24th, the Pet was kept going all night, and reached Bridlington the next noon, entering the harbour on the evening tide, notwithstanding a morning calm. Not at all a bad passage. One can gather what kind of sailor Hughes was from his remarks on Bridlington Harbour. "Bridlington quay, with all its defects, is a very convenient place, but with a gale of wind, especially from the N.E., if it is thick, keep the sea."

The Pet was headed north, and stayed for part of a Sunday at Whitby, bringing up in the Roads whilst the owner and his brother went to church. R.E.H. makes a disparaging remark or two on the indifferent attendance of the natives at divine service. He was no great stickler himself in the matter of Sunday observance, for he proceeded to sea the same afternoon, and just plugged along until he reached Leith,

going outside the Farne Islands, and, after being at sea three nights, entered his port at 4 p.m. He seems to have had very mixed weather, with the wind mainly fair from light to strong, so the passage was not a fast one. The speed of the yacht with the wind fair and all sail set may be judged from the statement that he easily reached past a pilot coble, and in two hours had "run her nearly out of sight." Perhaps seven and a half knots at the outside, probably only seven. The distance from Whitby to Leith is about 130 miles as sailed.

The rudder had given great trouble in former cruises, having been rehung no fewer than four times. It was again behaving badly during the voyage to Leith, and the indefatigable Editor remarks in a footnote: "The subject of jury or makeshift rudders is too little studied by yachtsmen." But an inspection at Leith showed that it was all right this time, though it was evidently very badly hung, and thumped a good deal.

On August 7th an early start was made for Inverness, which was reached on the 10th. Going up the Cromarty Firth in smooth water, the owner estimates that the yacht reached a speed of "nine knots or thereabouts"—a pardonable exaggeration or over-estimate. We always think our boats are going faster than they really are; this is quite a human frailty.

I suppose the Scots of those days were to south country eyes a wild and uncouth race. Hughes seems to have rather disliked them and their habits, but while recording the objectionable traits of manner he at the same time gives them a good character for politeness. "I cannot sufficiently admire the kindness and civility of the Scotch. Men, women and children are alike remarkable for the attention and good nature which they show to strangers." Those of us who in later days have covered the same ground can cordially endorse these words.

The Pet was taken through the Caledonian Canal, and made her way down the coast to Crinan. Passing through the canal to Ardrishaig, and continuing her voyage down Loch Fyne, coasting the east side of Arran, she was kept going until Peel, in the Isle of Man, was reached, after two nights at sea. A pilot was taken from Fort Wil-

---

10 No positive identification.

liam to Ardrishaig, but the rest of the journey was made without any help of this sort. In a very playful and humorous passage, Hughes, in the manner of the fifteenth century, endeavours to account for the tailless cats of Manxland. " ' It be sayde of somme,' saith an old volume that

rising gale. Leaving this port with a foul wind the Pet was turned to windward for some days. Off the Land's End they met with rough weather, and shipped a sea which almost smothered the little vessel. I cannot think that there was a lot of wind, as the whole mainsail was carried at the

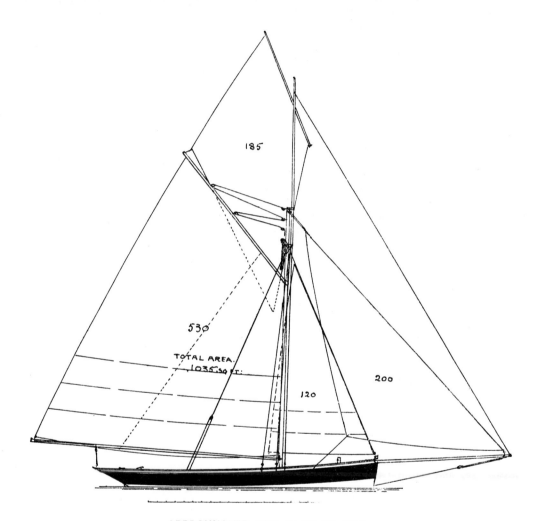

APPROXIMATE SAIL PLAN OF PET

I have consulted, ' that ye cattes of this Yslonde have none tayles upon theyre backes, whereof hystoryans doe alledge dyverse, curyose and exquysyte reasons,' " etc., etc. A very entertaining skit, anticipating the humour of Anatole France when he writes in mediæval French, but far too long to quote here.

His next port was Holyhead, thence to Milford Haven, which was reached in a

time, but there was evidently a lot of sea, and the yacht was hard pressed to take her over the hot lee-going tide. Abreast the Runnel Stone sail was reduced to close reefs, but with a beam sea they had a most unholy time. However, once clear of the Stone it was " down mainsail and up try-sail," and with double reefed foresail and storm jib she went up Channel, " swaggering away as bold as a man-o'-war," and the author says, " In running through the Lizard Race on this occasion, we found the Pet dryer than she had been all day." And so the little ship reached Falmouth, going on to Plymouth next day in a very curious rig-out of sails, which Hughes describes as follows : " I believe it was blowing hard from the westward, but we had had enough of foul weather, and determining to make fine weather of it *nolens volens,* we set the gaff topsail and balloon jib, stuck up half-a-dozen wet jibs for studding sails, and the trysail for a squaresail, and made her smoke through it. I never saw the Pet run as she did that day : a steamer which left Falmouth with us could not shake us off at all, and entered the Sound but a very short distance ahead." I daresay there may have been some slow steamers about in those days, but as Hughes carefully omits to state the time between leaving Falmouth and entering Plymouth Sound, we must draw our own conclusions as to the rate of sailing. After Plymouth he seems to have taken smaller bites at his passages, but eventually reaches his home port, Lowestoft, after a series of " dustings," in safety somewhere about the middle of September.

It has not been possible to give many extracts from the log of this excellent cruise, which was considered a very remarkable performance by the yachtsmen of those days. But it is quite obvious that Hughes did not think it in any way extraordinary. He believed that the Pet would go anywhere with decent handling and reasonable weather, and seemed actually to prefer being at sea at night to lying in a harbour. The capable crew he carried of course enabled him to pursue this course with advantage, and without courting overwork and exhaustion. Correspondence naturally arose, and one gentleman under the *nom de plume* of " Briny Deep," was very severe on the rashness of the owner of the Pet. He wrote : " Few persons are aware of the risks incurred by the owners and crews of small yachts in their perilous adventures at sea in six and eight ton vessels. It would be far safer to undertake voyages of the kind in good open sea boats of half the tonnage of these little lumps of wood and iron, which are as sluggish in a tumbling sea as a watercart. A twenty-feet ship's launch would ride over the waves like a cork, whilst the heavily ballasted little clipper would go down stern first," and so on. This, however, did not deter the owner of Pet from undertaking a still longer and more daring cruise to the Baltic in 1854, in order to see the actual fighting in that region, where the British Fleet was attacking the Russian fortifications on the islands and at Bomarsund and other places.

The outward voyage was uneventful, and was favoured by fair winds and fine weather most of the time. The yacht left Lowestoft on July 14th, and arrived at the mouth of the Eider on the 18th at daybreak, going up the river to Tonning on the next tide in charge of a pilot. This was, of course, before the days of the Kiel Canal, but the Eider was canalized, and after passing through Rendsburg it entered Kiel Bay at practically the same point as the modern canal does.

Hughes' remarks on the feeling of the inhabitants of Schleswig-Holstein towards the Danes are interesting to-day, as they seem to show that there was even then much dislike of the Danes, and a strong desire to be under German rule. He says, speaking of the Danish garrison at Rendsburg, " They appeared very quiet well be-haved fellows, and did not seem to suffer at all from the devout hatred with which the ciitzens regarded them." But there is no doubt at all as to his own feelings towards the two nations. In one place he remarks, " Often when travelling in Germany I have looked in vain among the flat-sided, broad-footed, wide-faced low caste natives for some trace of kindred race and origin with ourselves; but in Denmark you are con-stantly encountered by groups who would pass muster anywhere for the Anderson girls or the Johnsons, and upon inquiry they will probably prove to be the Johannsen girls or the Andersens. Besides this it must be confessed that the Danes appear to share with ourselves that peculiar pro-pensity for washing their hands and faces —doubtless an absurd and insular prejudice —from which our other continental neigh-bours—we must do them the justice to say —are generally exempt."

This was said in reference to the inhab-itants of Copenhagen, where he had been compelled to call in order to land one of his paid hands, who was very ill, but was most kindly received into the hospital. On

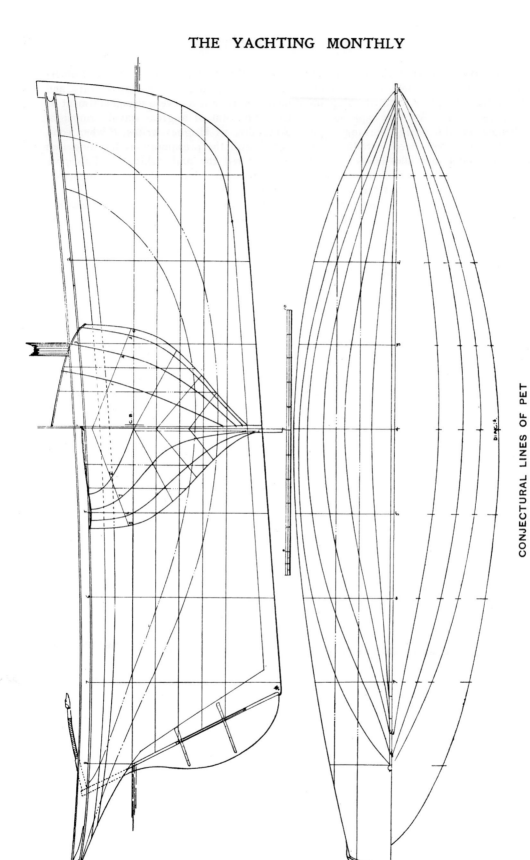

CONJECTURAL LINES OF PET

L.O.A., 33 ft.; L.W.L., 28 ft.; Beam, 8.5 ft.; Draught, 6.5 ft.; Displacement, 12 tons.

the voyage from Kiel, which, though nominally Danish, he labels as a " great, stupid hot German place," an amusing incident is recorded. As they approached close hauled in the early morning light breeze, the pilot had been extolling the speed of one of the new pilot boats. " She sail by we like *dat*," said he, drawing one horny paw rapidly past the other.

" Presently a taut white-pointed topsail came stealing out from the dark shadow of the land.

'What craft's that, pilot?'

' Dat is the new pilot boat, by God,' answered our Palinurus. ' Now you shall see her go !' Here a pause ensued, during which the pilot became more and more fidgetty, till at last he sang out, ' De pilot boat do drop astern, by God !' And so she did; but they are fine, powerful boats notwithstanding."

One has no difficulty in discerning that in light weather, with her enormous sail plan, the Pet could ghost along, but even in fresh winds she was generally far faster and more weatherly than any of the local boats, pilot boats and all, that were fallen in with.

Having seen the sick member of his crew comfortably ensconced in the hospital at Copenhagen, and shipped a Dutch paid hand in his place ("a very sorry exchange," says Hughes), the Pet was headed for Slitehamm in the island of Gothland, which was reached after a passage of two days. Here he fraternized with some officers of the Swedish navy, and after being nailed in harbour for several days by bad weather put to sea again in a fresh foul wind, and eventually, after a bit of a dusting, found part of the British fleet anchored in Faro Sound. Here he was informed that active operations were in progress at the Aland Islands, and after provisioning the little ship with " a quarter of an unhappy cow, a stock of biscuits, hams and cheese, and a dozen or two of glorious golden sherry, which had been originally destined for the cellars of the Czar," left with fine weather and a fair wind for Led Sound, where the Allied Fleets were anchored, and the adventurous Pet and her crew reached their goal amongst them.

Hughes, like many another, was grievously disappointed at the inaction and futility of the whole of the fighting on the part of the Allies. The capture of Bomarsund was the only successful military and naval effort, but a severe disappointment as the sum total of a great naval campaign, especially, as Hughes writes, " when compared with the presumptuous boasting that inaugurated the undertaking." This poor result was not due to the incompetence of the Fleet, but was caused solely by the inactivity and timidity of the home Government. *Absit omen* !

As Hughes gives no dates, or only very rarely, I cannot gather exactly how long he stayed on the scene of action, but as he arrived back at Copenhagen on August 31, his stay was less than a month.

The Pet now fell in with bad weather, and after making a start had to put into Fladstrand for shelter and to land one of his paid hands in the incipient stages of cholera. It was not until September 9th that the little ship was able to leave for her long stormy passage of 480 miles across the North Sea short-handed, and with the added anxiety of the illness of Lieut. Hughes, who became unable to assist in the working of the ship, which set forth with a pleasant westerly breeze in company with a squadron of small Swedish and Norwegian cutters. Hughes remarks, " It is extraordinary what wretched craft these north country cutters are. In plying to windward I do not believe there is one of them that can go nearer than 6 or 6½ points allowing for leeway; all the evening and all the night we continued overhauling them one after the other until on Sunday I think we had got forty or fifty under our lee."

By pluckily sticking at it, the Scaw was passed, but before long the sea grew heavy, the wind increased, and a hard struggle began. " Having gained an offing, I got down my mainsail and stowed him for the last time, set the trysail, close-reefed the boltsprit—(Hughes always writes *boltsprit*) —and with storm jib and reefed foresail sent her at it once more."

Now became apparent the value of length, weight, and an easy section in this thrash to windward over a stormy sea. " I continued to feel my way across by the lead, keeping the little cutter on that tack on which she looked best up for her port. The little ship behaved admirably, weathering and forereaching everything we fell in with in a surprising manner, fine clipper barques and schooners that certainly would have forereached us in fine weather now fell to leeward and dropped astern, and that, too, so rapidly that I could scarcely believe my eyes. But it was hard and anxious work—on one occasion I stood nineteen hours at the helm, and then only abandoned it when it came on to blow such a gale that we were compelled to heave to.''

The little Pet must have been a perfect wonder hove to. The plucky owner took refuge below from the bitter gale, the vessel being then under double-reefed trysail and storm jib and the cockpit battened down. Under these strenuous circumstances the tough crew sat down to a meal of ' a strong brew of hot tea, toast and Welsh rabbit to console us, one hand looking up every minute to see all right.'' She was hove to all night until noon next day. The next night at midnight sailing to windward was impossible, and again she was put head to wind.

They spoke two vessels to compare reckonings, having no quadrant on board. It would have been impossible to have made effective use of one if it had been there. But the extreme weatherliness of the Pet is made clear by the following remark : '' On both these occasions we found it necessary to come about a long distance on their lee quarter, and by the time we came up to them their leeway was sure to bring them down within hail of us.''

Nine days were thus spent, battling against head winds and seas, and on the ninth evening two welcome shore lights were sighted. On the tenth day land was seen, and the little ship reached her home port, Lowestoft, '' after zigzagging across 480 miles of sea, never for an hour on our true course, after being thrice hove to, on

one occasion for 15 hours, the first landfall was the very spot for which we were bound. We heard afterwards that the Maria, schooner, which had been our neighbour in the roadstead at Copenhagen, had foundered in the North Sea, and four other vessels had perished during our passage on the same voyage.''

The Rev. Robert Edgar Hughes, M.A., deserves remembrance in the minds of modern day cruisers. What a fine sailor he was, and what a shipmate to have been at sea with ! I think it is possible to improve on the Pet in some few ways, but after all most of the essentials of a good cruiser are contained in her old fashioned carcase.

It is comforting to think that now, when we live in what are often called degenerate times, Hughes' pluck and seamanship are more common than in his own day. Claud Worth[11] has done as much—perhaps more—with an amateur crew in a very similar ship, and think of the marvellous single-handed cruise in the Bay of Biscay by A. Stanley Williams in Ben-y-gloe,[12] a much smaller vessel. The journals of the Royal Cruising Club contain records of many similar cruises, and at this moment of writing what countless thousands of Britons—at sea, in the trenches of Flanders, and on the blood-drenched shores of the Gallipoli Peninsula, are proving that the old spirit lives and shines with a tear-compelling lustre, as bright and searching as in any period of our thousand years' history. *Laus Deo* !

NOTE.—*The account of the Baltic Cruise is contained in '' Two Summer Cruises with the Baltic Fleet in 1854-5,'' published by Smith Elder & Co., 65 Cornhill, now long out of print.*

For footnotes, see page 218.

EPILOGUE

In June of 1916 a relative of Rev Hughes responded with further information about the Hughes family.

THE CRUISE OF THE PET AND THE REV ROBERT HUGHES

Sir, With regard to your interesting article in the April number on the cruise of the Pet, *the author says, 'nothing is known of the owner, the Rev Robert Hughes'. Being a near relative, I can give the following information: He was the son of the Rev Sir Thomas Collingwood Hughes, ninth Bart. (see Burke), whose ancestors helped to make naval history both before and after 1773, when Richard Hughes was commissioner of Portsmouth Dockyard and received his baronetage for entertaining George III. Another Sir Richard was second in command and Admiral of the Red at the relief of Gibraltar, and captured the Soletairo. He was honourably mentioned by Mahan in his famous book. Robert Hughes' brother, his companion in the* Pet, *Capt J D Hughes, afterwards built the* Vanguard *at Ratsey's, one of the most successful 60-tonners in the squadron (1866) and subsequently owned several smaller racers and cruisers. He was one of the founders of the Y R A and acted as secretary until, at his suggestion, Mr. Dixon Kemp was appointed. All Capt Hughes' children have followed 'the call of the blood' and raced regularly on the Solent since 1866, Mrs Shenley, his daughter, having owned seven racers in various classes and won some hundreds of prizes. Capt Hughes would like to say that the 'conjectural lines' of* Pet *are wonderfully correct.*

B S Hughes

Strange replied in the July issue, adding information he had received on the ultimate fate of *Pet*.

THE REV ROBERT HUGHES AND THE PET

Sir, Will you allow me to express my thanks to Miss Hughes for so very kindly furnishing further information concerning the Rev Robert Hughes, which I am extremely glad to learn. He probably made more cruises before parting with Pet *and it is a pity that no other records of his voyages and his life have been granted us. Truly the salt in the Hughes blood has not lost its savour, as is very evident when we remember the keenness shown by Mrs Shenley and Miss Hughes in their racing on the Solent.*

It may perhaps be worth recording that my friend Mr Archibald Dickie of Tarbert was the last owner of Pet. *He writes to say that he bought her at Fairlie some thirty-three years ago, to break up, after she had been shored up on the beach for a long time. He corroborates Capt Hughes' opinion as to the correctness of the conjectural lines and adds that he much admired the shape of the old vessel.*

If boats could only speak, what a tale this wonderful little craft could tell!

Albert Strange

---

11 Details of the various yachts Worth owned, and the cruises he made in them, are to be found in his book *Yacht Cruising.*
12 *Ben-y-gloe* was built in 1901 to the design of G U Laws. LOA 31ft, LWL 23ft (approx.), beam 7ft, draught 5ft – at times rigged as ketch and yawl – with a canoe-stern. In her, A. Stanley Williams cruised single-handed from Southampton to the Gironde in the Bay of Biscay and back in 1914, a distance of some 1,000 miles.

CHAPTER 5

A Winter's Tale

The Yachting Monthly, May and June 1914

Humber Yawl Club Yearbook, 1897

A WINTER'S TALE

This and the account presented in Chapter 6 were written some years after the events took place. 'A Winter's Tale' was published in *The Yachting Monthly,* in May and June, 1914, 32 years after the cruise it describes. It is infused with that blend of retrospection and freshness which indicate that these 140 miles and 10 days in a tired, unsuitable, old boat and bitter weather were, indeed, as Strange put it, 'deeply graven upon the mind' and, as we would now say, a formative experience.

'A Winter's Tale' describes a passage which many would consider somewhat ill-conceived and dangerous by modern standards, but is characterized by a youthful recklessness which is, hopefully, no less in evidence today than in 1882, when Albert Strange was 27 years old.

Of today's amateur coastal yachtsmen who have experienced sailing in a snow storm, few would contemplate it without the efficiency of modern clothing and heating, and the reliability of auxiliary power. Yet the undertaking of a winter passage in such an under-equipped boat must have seemed less drastic in an age when thousands of small craft, many of them undecked, were worked year round to earn a living for their crews.

The year 1882 was something of a turning point in Albert Strange's life. Having returned from eight months of cruising and sketching in *Quest,* he was to marry Julia Louise Woolard in August and take a responsible post as headmaster of the new Scarborough School of Art in the following month. Whether the news which 'compelled' Strange's return to England in March was connected with his appointment to the post, it is not possible to say, nor how long he might otherwise have spent in France.

It is to be regretted that it has not been possible to establish the identity of the Poet.

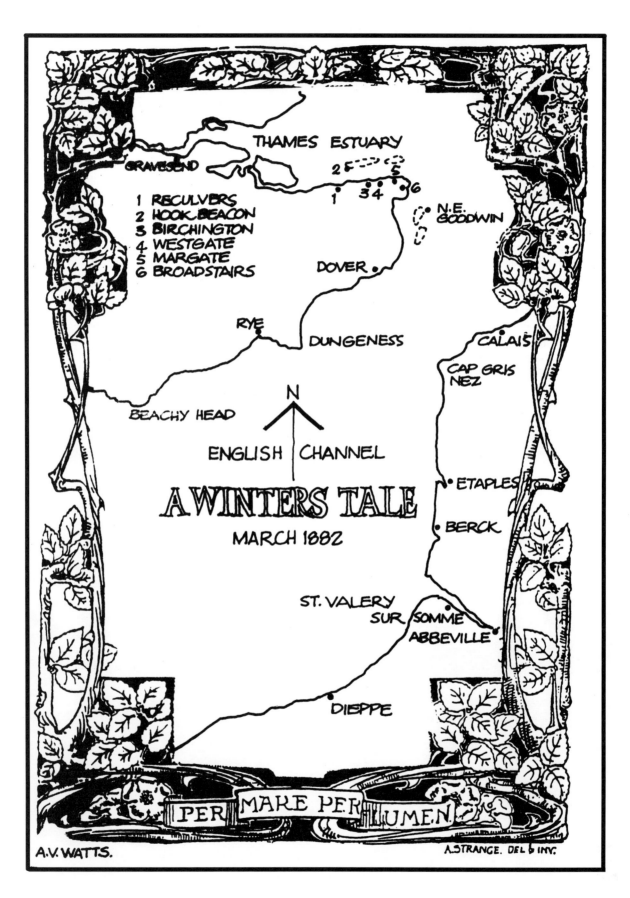

THAMES ESTUARY

GRAVESEND

2
5
1
3 4
6

N.E.
GOODWIN

1 RECULVERS
2 HOOK BEACON
3 BIRCHINGTON
4 WESTGATE
5 MARGATE
6 BROADSTAIRS

DOVER

RYE

DUNGENESS

CALAIS

CAP GRIS
NEZ

N

BEACHY HEAD

ENGLISH CHANNEL

A WINTERS TALE

MARCH 1882

ETAPLES

BERCK

ST. VALERY
SUR SOMME
ABBEVILLE

DIEPPE

PER MARE PER LUMEN

A.V. WATTS.

A.STRANGE. DEL & INV.

Outside the Somme.

INTRODUCTION.

*T*O every man who has adventured, and who has taken his chances with
wind and wave, this fragment of a youthful cruise is dedicated. It has been
gathered from old log books, old sketch books, and old memories deeply
graven upon the mind. Perhaps some of the joys then felt may still be conveyed
through its inadequate words to the heart of the cruising yachtsman, who will, the
writer is sure, extend his indulgent sympathy to the effort here made to prove how
great is the joy of a struggle against the spite of calm, fog, storm and snow, when
the men and the boat are in the end victorious.

A Winter's Tale.

BY

ALBERT STRANGE.

This is a tale of the past—long spent—with the long days of youth,
Days that were jewels ashine on the glistening circlet of Life;
Full to the brim was the cup, full to the brim was the heart,
Brimming o'er with the strong wine of hope and the passion of strife.

Many the tides that have flowed; ah, many the moons that have waned,
Many the ships that have sailed but never returned to the shore.
Tired are we now of the strife, weary are we of the toil,
Here is the haven at hand, and we may adventure no more.

Here with the slumbering hulks, sea-worn and shattered and spent,
Here shall we moor by their sides, here with the call of the sea
Sounding aloud in our hearts, and the cry of the gull on the breeze;
Calling up dreams of the past—deep dreams of the days that were free.

PART I.

WHEN Youth, and Hope, and a Poet are one's shipmates, many shortcomings of the ship may be condoned. Even a "converted" boat may bring some measure of satisfaction, especially in fine weather and smooth water; for under such conditions almost anything will serve as a home. But when winter has come, and winds and seas are rough, then contentment dies a quick but painful death, and the naked inadequacy of such a craft stands revealed in all its menacing hatefulness.

The original Quest (*née* Great Harry) was a converted Government cutter. Some poor fool had given her a deck of portentous thickness and weight, a light iron keel, a low cabin top, and a grossly superfluous cutter rig, under which Rumour said she had won prizes. About a foot or so had been added (but not properly secured), to her original topsides, and with some internal ballast of under a ton and the small iron keel of unknown weight, she drew rather less than three feet, not a big draught for her 26ft. of length and 7ft. of beam. A few slight floors had been fitted across her bottom to take the iron keel bolts, but no other strengthening had been given her, so the hull, being clench-built, "worked like a basket."

I have heard that this peculiarity is conducive to speed, but it did not make Quest fast, and it certainly did not add to her comfort or dryness below. To give the poor creature her due, she could get out of her own way on a reach or running. But on a wind! Heavens! how slow she was, and how often she mis-stayed! It was then that you felt how much too much the £20 she cost us was, and what a hollow fraud a converted boat can be.

In this apology for a yacht, now under yawl rig, the Poet and I had spent some eight months or so crawling about the Thames Estuary, the South Coast of England, and the North Coast of France from Calais to Dieppe. Early March found us at S. Valery-sur-Somme with sailing orders for home, after having spent December and January at Abbeville with our boat garlanded with hurdles to keep the ice floes from staving in her bows.

The *quai* at Abbeville was infested with rats—large rats—and we were never short of sport, for they boarded us in bands and droves each night, hard driven by hunger and cold. It was no uncommon thing to see one or two foraging over the cabin floor after we had turned in and all was quiet. We got used to them, but we never became friends in the fullest sense of the word.

We also got used to the ever-present row of sightseers who came, especially on market days, to gaze upon the mad Englishmen who had come from *Amerique* (?) in this crazy little vessel. They would stand and jabber about us for hours, and if we were below would kick small stones from the *quai* on to the cabin top in order to bring us out. The Poet, true to the nature of his tribe, being of an irascible temper, would hurl curses at them, at which they would smile. The other man said little, but would sprinkle them with water from the mop, which pleased them much less, but was probably taken as part of the performance. Then Pierre Dieudonné, who kept the *café* near by, hearing the row, would come out and explain us to them in Picard, after which explanation they would retire to the *café*, and we to our cabin.

Pierre must have hinted that we were *dangerous* lunatics, for if we put on a desperate air and climbed the quay ladder quickly, the crowd would melt away amazingly fast. The little boys of Abbeville (who also threw stones) would, in our walks about the town, follow at a respectful distance; but if we stopped suddenly, or turned round, they would bolt up the nearest alley, casting fearful glances behind as they ran.

Abbeville, however, had on the whole, been most hospitable. The skipper obtained permission from a gracious *maire* or *sous prefet,* to sketch where he liked. The Poet was made free of the big libraries at the *Musée* and at the Town Hall, and was freely permitted to handle precious manuscripts written when Abbeville was young. The junior officers of the garrison were kind to us, and would come and sit in our cramped little cabin, taking lessons in English conversation, listening to the Poet's sonnets, and turning over the skipper's sketches. We dined together, were made free of their *cercle,* and were, in fact, completely spoiled by kindness.

But they itched to go for a sail in the yacht, even in that bitter weather. This the skipper always deprecated, knowing the creature's unhandiness. However, when he once left the Poet to himself for a few days whilst he went up to St. Riquier to paint, they got their wish, and persuaded the Poet to cast off and take them for a sail on the rather tricky river, then running in flood. The poor Poet knew less than nothing of sailing; he never even at the very end of the cruise learned the names and belaying places of the halliards, and when the skipper returned home from St. Riquier he was horrified to find the little vessel minus bowsprit, topmast and mizenmast—a clean

sweep! Between them, the Poet and his soldier crew had sailed bang into the railway bridge, and the current had done the rest.

But they gave us new spars, and laughed at the Poet, whom everyone loved, especially the ladies. In truth, there was one, a dark-eyed *rentière fort brune,* with a handsome *dot,* who seemed specially attracted by him. Pierre Dieudonné, who knew everybody's business and every girl's fortune, was always urging him to lay siege and win the lady, who had, it seems, 300,000 francs, a hasty temper, and an artificial leg. The Poet thought there was just one defect too many, and remained coy.

Suddenly news came from England that compelled us to return, and it was a cold, dark day in early March when our retreat from the charming old French town and its people was made. A light breeze fanned us slowly down the eight miles of canalized river, which ends in the great locks at St. Valery-sur-Somme. These open on to the tidal harbour and the vast estuary of sand which lies beyonds its mouth.

Up the harbour every flood tide runs a big wave, a part of the " bore " which races up the main river off which the harbour is built. This wave varies in height according to the state of the weather as well as the nature of the tides. At springs it is naturally bigger, at neaps scarcely perceptible. It does you no harm as a rule, but if your boat is on the hard near the locks it may shift and bump her violently, as it did Quest, who was then being scrubbed before making our Channel passage. She wanted this scrub badly, for though there was no marine growth of barnacles her bottom was covered from keel to waterline with a curious brown hairy crop of weed, which made sailing almost impossible. It took us two tides to remove this, and even then she was very rough, but we hoped that she would manage to get home without further attention. The scrubbing over, we provisioned, and then shifted to the lower part of the harbour to await the night ebb, on which we should make our start for home. High water was about 11.30 p.m., and our plan was to dine on shore at nightfall at the " Colonne-de-Bronze," an auberge which possessed an excellent cook, and was owned by a simple-minded proprietor of non-predaceous habits; and so make a fittingly ceremonious farewell from the land in which we had found such kind hospitality for so many months past.

I do not know whether the "Colonne de Bronze" still exists. It may by now perchance have blossomed out into a real expensive hotel; but the way in which Philomêne, the buxom cook, prepared grey mullet, soles, mussels and flatfish, was so surprisingly good that if ever splendid deeds

bottles should help to keep out the cold, which was intense, and see us across the Channel, and we were not far out in our reckoning.

After dinner came the farewells. The Poet sought out Philomêne in her kitchen, his adieux were protracted, and accompanied by what sounded like the smashing of crockery. Then he emerged hastily, pursued by Philomêne's mistress. After that

Abbeville.

have their reward here below, hers should not escape praise. Her cooking was really lyrical, and should have been a monument to the fame of the inn. The white wine at 1.25fr. the bottle was almost equal to the perfection of the cooking. According to the Poet, it was Nectar, but the skipper contented himself by describing it as a non-intoxicating, but highly exhilarating beverage of most uplifting properties. Such, you see, is the difference between the commonplace and the poetic mind when dealing with concrete facts. We thought one dozen

we dragged our dinghy down the steep slipway, got on board, shortened chain, and began to prepare the vessel for sea.

We had a small coal stove fitted at Abbeville, and this had to be secured by stays of wire to the mast and to any available projection. Then the canvasses and the sketches, the result of the skipper's own special toil, were wrapped up in waterproof and fixed under the fore deck, the only really dry part of the ship. The Poet's immortal manuscript, his poems, and that still unpublished work, "A Mediæval

River," upon which he had for so long been engaged, were locked in his big tin box, and so made secure from destruction, and then we were ready for sea. The 120 miles of grey, wintry sea seemed a long way to travel in a crazy, leaky craft at this inclement season, and had we been less youthful we should probably have left the boat to rot where she was, and betaken ourselves and our belongings home by train and steamboat. But I think we both felt it incumbent on us to bring the boat home too, and so completely round off the eight months of adventure and enjoyment. Besides, had not some croakers prophesied when we set off that neither we nor the boat would ever see England again?

There were no leading or other lights to guide us across the sands, except a wee faint harbour light that soon became invisible in the misty night. The channels were winding and shifting, and the only way open to us was to set a course somewhere about N.W. and trust to luck to carry us over the shoal places and clear of the half-submerged buoys that the seven-knot ebb would soon be riding under. So it was up anchor on the short slack at high water, and get away under all sail, so as to clear the estuary and reach the deep water of the Channel as soon as possible.

It was now that we discovered the loss of our topsail! The Poet had been engaged in clearing out our sail locker whilst we were scrubbing on the hard, and vowed that he had replaced everything. Our big jib, and the spinnaker (which we hardly ever used) were there, but, alas! no topsail was forthcoming. This was a serious loss, as it made a considerable difference to the speed of our boat in fine weather, and we hoped for light winds, as the glass was high and a severe frost was in progress. But it was gone, and there was no help for it; some predatory fisherman may have taken it, or, what was quite as likely, the Poet may have neglected to stow it away, and when the " bore " hustled the boat about it may have rolled overboard.

So we set all sail possible, and with a light E. wind got our anchor on the slack and stole out into the darkness of the night on the first of the ebb. We had some seven miles of sand banks to get over before we reached the Channel, and we trusted to our luck and to that special Providence which

looks after fools, drunkards and children, to take us clear of the high sands, which would soon begin to show with the very rapid fall of the water so remarkable in the Bay of Somme. But though they might show we should not see them until we touched, so the first hour of our voyage was full of anxiety and frequent soundings. Presently we heard a faint roaring sound, which rapidly grew louder, and suddenly we were rushed past a foaming hillock of water which boiled over one of the N.W. pass-buoys. Another and another rushed by, then a comparatively long interval elapsed before we drew near any more. The darkness made the sound of the foaming ebb seem louder; the boat appeared to tear along whenever we passed a buoy, but felt almost stationary when they were astern. It was a time of tense anxiety, during which neither of us spoke an unnecessary word. Suddenly we found ourselves in the midst of a space of sea covered with short pointed waves, which clapped and danced all around us, and then a slow, gentle heave told us we were clear of the Somme sandbanks, close by the big outer buoy, and on the free Channel sea.

It was now nearly 1 a.m., and the wind still kept very light from the E., but the mist grew a little thinner overhead, where the moon should have been, and gave us hopes that when daybreak came we should be able to see the land, if we kept a N.N.E. course. So, after a fair offing was made, this course was set, which would keep us along the land, and at any rate we might get a sight of some fishermen out from Etaples or Berck, and thus learn our position. Moreover we should be well out of the way of all steamboats, and if it fell absolutely calm we could anchor in reasonable soundings with our long kedge warp.

So all seemed well, and the Poet was ordered to turn in, which order he obeyed after uncorking a bottle of the Nectar and drinking adieu to France, in which libation the skipper also took part.

That seemed to be a long, long night, the little ship creeping slowly along, rolling gently in the swell, her sails stiff with the frost, which whitened deck and spars, and hung in filmy threads from the ropes. I doubt if she held her own against the ebb, which hereabouts runs with speed and vast irregularity of direction. When the first intimation of daybreak made itself felt rather than seen, through the mist, the Poet was routed out to get breakfast, the ship was again pumped dry, and soundings were taken; 13 fathoms and about low water seemed all right, and so the course was kept by the Poet whilst the skipper took a little nap after breakfast. The little nap seemed a very short one to the tired skipper, when he was called out by the Poet, who declared

took off his oilskin and his comforter and went below to warmth and a higher degree of luxury than that existing in the cockpit.

By noon—a rough estimate derived chiefly from tidal calculations and frequent soundings, made us 16 miles N.N.E. of the mouth of the Somme—the distance off the land being unknown, was thought to be about 6 miles. And so passed that day, we scarcely moving on our course, and no sight or

St. Valery-sur-Somme.

that he was lost, and that the boat would not do anything, whichever way he put the tiller. A peep outside revealed the sad fact that she was head to what little wind there was, mere catspaws; her head sails aback, and practically hove-to. A cast of the lead gave 17 fathoms, which seemed to show that she had gone considerably off her true course. No land was in sight, the mist was thicker than ever, and we were in a little circle of sea and fog, cut off from sight or sound of anything. This gave the Poet several ideas for sonnets, and he thereupon

sound of any living thing except a few duck. The next night seemed desperately long, and was nearly calm; little tricky flaws of wind from the E. just gave steerage way. Had the skipper been as wise as he thinks he now is, he would have brought up for a few hours for a real sleep. But he stuck it out until dawn, when utter weariness compelled him to once more relinquish the tiller to the Poet, with strict injunctions and commands as to what to do and how to keep the course. About three hours after turning in he was awakened by the Poet, who announced more wind, and the boat certainly was moving along a little better. But at noon our little breeze left us almost becalmed. Early in the afternoon we fell in with a fishing boat, and the Poet was sent away in the dinghy to inquire as to our position. He came back with a bucketful of fine whiting and the information that we were gradually losing our hold of the land, caused as much by the tides as by the leewardly propensities of the boat.

A dash of ice-cold sea water across his face would keep him going for half an hour, and then again utter weariness would overtake him. Oh, for only an hour's sleep! but it would have been madness to have put the Poet in charge. So he fought against fatigue with more cold water, stood up and steered, moved about the little cockpit and longed for the day. At last the Poet had to come on duty and wake the poor skipper whenever he nodded, for he began to hear

Bay of the Somme at St. Valery—Low Water.

When night again fell no lights were visible, no sound of fog signals was heard, and we again sailed through deep darkness. The breeze was a little fresher, and the yacht was sailed a point or two closer in the hope that we might perchance get a glimpse of Grisnez Light, and to keep out of the way of traffic. But no light was seen, nor was any sound of fog signal heard as the boat, heeling a little more, went rustling along in the gloom. It was most difficult to keep awake as the night wore on, and several times the skipper caught himself fast asleep with his arm still on the tiller.

church bells, bands of music, sirens calling, and the hundred odd sounds that a dead tired and anxious man hears in the night at sea.

Then, as he steered on by instinct, blind and stupid, the Poet shook him and said he could hear surf! *Surf!* This was enough to wake a dead man. Yes; there it was—to leeward, sure enough—we were close to the back of the Goodwins, and the mist still thick. We could still luff a little, and kept our wind as well as we could until faint signs of dawn appeared, and with them the fog lifted, and right ahead was a ketch standing N.W.—just silhouetted against the horizon.

Now we knew where we were—nearing the N.E. Goodwin buoy. Joy! The tide was still fair, and the wind had a little more heart in it. Ports and a good sleep were near at hand, with any luck. Another, and this time a complete, sluice down with sea water, woke the skipper up, and he shaped a course for Broadstairs, which he knew not, but which would soon be in sight, and we might save water into that place. We rounded the little pier with 3ft. to spare, pulled down the sails, made fast, and then, leaving the Poet to explain things to any Coastguard who might appear, the tired skipper, after one good pull at the Nectar, collapsed in his berth and slept the sleep of the just, all standing, oilies and all!

A Winter's Tale.

BY

ALBERT STRANGE.

II.

BROADSTAIRS is not an ideal harbour. It is very small—just a bight in the cliffs with a wooden breakwater which stretches an elbow from the north side, and forms a curve within which a very few small craft could lay. There is a good deal of range at all times, but with the wind in from S.E. to S., it must be a fine place to lose your vessel in.

Outside, the ridges of rock stand high above the low water mark. They probably do break a little of the sea, but not much.

After a solid fourteen hours' sleep the skipper arose, washed himself, and had a meal—supper I suppose it must be called, as it was about ten o'clock. The Poet was ashore, and presently returned with a jar of beer, some shag tobacco, and a local paper. These things had an exquisite flavour of home about them, as had also the parcel of bloaters he produced from his pea jacket pocket. He reported a thaw, and said that the local fishermen, or whatever they were, predicted that we should have to clear out next tide, if the wind came S., as with the thaw it might well do. However, though we were afloat, the tide would not serve until next morning, and between now and then we should spend a good few hours on the ground.

The joy of being in a harbour had filled the Poet with immense excitement. We were fairly home now we thought, another tide or two would see us up to the Gravesend Canal Basin, where we intended to leave the boat until we might find a buyer for the old trap. So a good part of that night was spent most harmoniously discussing plans for the future, for we were both going to settle down to pretty steady work, and make

our fortunes before we were many years older. "If youth but knew," had not then been written, but even if it had we were both too young and hopeful to have the least doubt about our magnificent careers.

Next morning broke with a cold S. wind, looking black and as if snow might be imminent. Before we were well afloat the old packet began to bump and shake and work like a basket, and the fishermen advised a retreat to Margate, where some of their boats were to be taken. There wasn't much wind, but it didn't look nice at all, and the range was quite uncomfortable. So we said we would go too. But the first thing was to get out of our sheltered elbow. There wasn't room to sail out, and we couldn't pull out, so we had to bargain with the fishermen to run away a long warp to a hauling post on the rocks on the other side of the little bay. They charged 7s. 6d. for this, at which we coughed a good deal and felt that we really had got home at last and no mistake. Our little dinghy was too small for the job, and our warp was only 30 fathoms, whereas the post was twice that distance away. So we paid and left, being able to carry whole mainsail with second jib.

It didn't take long to get to Margate, and we luffed into the mouth of the harbour, half inclined to stay, yet eager to go on. The wind, if it held, would just let us lay to Whitstable, but we couldn't hope to save our tide there. We *ought* to have stopped, but knowing what a shabby spot Margate is with the wind at N.W. (there was no telling that it wouldn't fly there before night), we bought loaves and meat and went on along the land, the skipper full of misgivings and the Poet complaining of having to pump the beastly boat out twice in four hours. As a warning the wind freshened a bit, and the old thing began to lay down and soak water through her topsides at a good rate—hence the Poet's tears. Pumping was always his task, it being the only job he could do on deck. And it had grown mortally cold and black. Still, on we staggered, going sideways as well as ahead, past the Hook, bundling along with more sail than we should have had on her, but doing our best to get along home.

It was about 4 p.m. and we were off Reculvers (a long way off) when without the least warning except a few flakes of snow, a heavy squall struck the yacht and laid her down on her beam ends. She was quite unmanageable until the headsheets were let go, when she slowly came up a little and lifted her lee decks out. The furious slatting tore the jib out of its bolt ropes, and the skipper clambered forward to let go the main halyards, when the sail jammed and had to be pulled down by the hoops. The poor Poet came out of the cabin, and as soon as he could make himself heard, said that the water was up to his berth below. We managed to wear her on to the other tack and three reefs were buttoned down, the mizen furled, and the reefed foresail set. Meanwhile it was snowing furiously and the wind had veered W.S.W. It took such a long time to get things decent that we had no idea where the boat had got to. All bearings were lost in the blinding snow. Again the Poet was called upon to pump, and we got a lot of water out of her, but not by any means all of it. Fearing that the boat would be blown off the land we kept her on the starboard tack for a good time and then bore up for where we supposed Margate to be. We could see nothing beyond a radius of 100 yards, and once off the wind the yacht tore along madly in unaccountably smooth water, considering what a power of wind there was.

Just as we were congratulating ourselves on having got out of a very nasty tight place, the yacht began to bump over the ground! The skipper bore up hard, but it was no use, and with a long grinding scrunch, she stopped and slewed. The stove, fortunately nearly cold, flew from its improvised lashings and the whole cabin was thrown forward, at least what there was that was moveable. Our three gallon water jar smashed against the stove. The tin box with the manuscripts was on top of that, with all the various things below that had fetched away. Shipwreck stared us in the face.

As it was just after high water the yacht soon began to hit her bilge against the chalk rock on which she was fixed, and the sight of her lee side bending up inside, every time she struck, was most instructive and curious. Plank and timbers seemed to lift a couple of inches, and at any rate it was not possible to deny the elastic powers of a clench built elm boat.

It looked as if we must be stove in very shortly, but a brilliant idea seized the skipper. Rushing out into the cockpit he dragged the sailbag out of its locker, rammed in big jib, spinnaker and everything it would hold, made a line fast top and bottom, took his clothes off hurriedly and going overboard on the lee side, with the Poet anxiously doing his best to help, got the bag, after some futile efforts, nicely under the spot where the bilge hit the rocks, and kept it there until she settled down a bit.

This, with the snowstorm going on, was naturally a perishing business, and the poor skipper on getting aboard, was at the last end. But the Poet rubbed him down, put him to bed, and administered half a bottle of chlorodyne, and said he would keep watch, and try to get the stove fixed up.

The boat did not make a lot of heel when the tide left her, and the stove being refixed, was again started, though with difficulty. Turpentine, being a part of the landscape painter's outfit, was fortunately available, and just as the skipper dozed off, almost stupefied by the chlorodyne, the last thing he remembers seeing was the Poet kneeling down and groping for bits of coal from the upset coal box forward.

It was against all rules for a skipper to turn in like this, but he was really played out for the time being. He fully intended to be out of his bunk long before next high water to relieve the Poet, and get things fixed up a bit. But the chlorodyne was too strong, and when he did wake up with a sickening taste of drugs in his mouth, he felt that something was wrong. The Poet was fast asleep, sitting in his bunk with his glasses on his nose, utterly tired out, as was the stove. Hastily dressing, the skipper looked out and found the snow still snowing and the wind still raging, but most alarming of all, the boat was afloat and slowly driving off. There was no knowing how long this had been going on. The time was 3.15 a.m. and no light visible, just a close circle of level driving snowflakes and cold spindrift, but little sea to speak of.

The lead showed 4½ fathoms, and without a moment's delay the anchor was let go. The wind being W.S.W. had still a bite of

was put into the bucket and brought into the cabin to melt. But the produce was only about half a pint, so more was scraped up until enough was got to half fill the kettle. All this time the poor Poet slept on, and was allowed to do so until the tea was made, and a cup was, well, not forced upon

A Sail Bag under her Bilge.

the land and we could not be very far off the soundings. After casting overboard the dregs of the chlorodyne, the skipper felt better, but longed for a cup of tea, and then it dawned on him that there was not a drop of water on board! We had lost it all when the jar smashed.

There was very little snow on deck, it had been blown or washed off as soon as it fell. But there was in one corner of the cockpit a tidy heap of it trampled hard and dirty looking. So, for want of better it

him, but eagerly accepted. The only plan now was to hold on until we could find out where we were. We thought we must be off Birchington, or perhaps Westgate, but it was quite impossible to say where. The skipper knew we could not be far off the land, because the boat rode fairly well and easily. She had 25 fathoms out, and a splendid anchor at the end of it, so we were fairly safe so long as the wind held in the quarter it was. We had no barometer on board, but both felt very uneasy, and the

Poet, who had maintained a dignified stomach so far, began to feel that he must lie down, as the motion of the boat at anchor was too trying. So he turned in and again soon slept, whilst the skipper, refreshed by his tea, and resolving to make the best of things, brought inboard the sodden sail-bag, which was still under her bilge, baled out the dinghy (wonderful dinghy that rode everything out and towed like a bladder),

the cable from the mast, and done all he could think of to defeat the forces bearing down on the poor little yacht.

When one has done one's best that is the end, and he could only wait through the interminable hours of darkness, the snow never ceasing, and the wind steadily increasing in violence.

Day came at last, but no cessation of the snow, and the wind grew into a gale.

She rode pretty well.

hoisted the riding light and wished (*how* he wished) for the dawn.

You see, it was enough to make him. A crazy, leaky boat, anchored God knew where; one sail blown away, and a torn clew in his mainsail; a mate willing and plucky enough, but useless in an emergency, and devoid of physical strength.

If this horrible snow would only clear away, the chief difficulty would cease to exist; but until he could see his way the skipper determined to ride it out as long as he could. The boat rode very well, and a good pump every two hours would keep her afloat for a long time. True, the bitts were very shaky, but he had put a spring on to

The comforting thing was that as the tide ebbed the lift certainly grew no worse. And it was not until twelve o'clock that the snow ceased sufficiently to show where we were. By some miracle we were only about three-quarters of a mile off Margate itself! on what must have been a projecting spur of sand or chalk, as there was about two fathoms at dead low water.

The skipper's joy was great, so great that he became boastful, and said he'd be blowed if he'd shift before he'd had a good feed. There would be plenty of water in to Margate for long enough yet. And feed he did, on some steak, washed down by some of the still remaining Nectar. The

Poet roused himself and tried to join in, but made quite a poor attempt to share in the little feast.

Let this be a warning to all boastful sailors. No sooner was the dinner eaten and the things cleared away than the wind veered to W. and blew a veritable hurricane. No canvas we had would have stood we had to give up, and went below wet through, meaning to try again later on, for to slip and leave anchor and thirty fathoms of good chain wasn't then to be thought of.

And as we sat and waited in our little cabin, now so uncomfortable from the motion that the only rest obtainable was by

We lay alongside the barge.

a moment, and so we had to ride it out in a rapidly increasing sea, the tops of the waves being blown in spindrift from the bows right aft. It blew like this for a couple of hours, and then let up a little. At once the crew attempted to get the anchor. They pulled and strove in vain. The Poet could not pull at any time; now, on the pitching bows, weak from sickness and want of food, he was no good at all. Two strong men might have got it aboard; a man and a half were no use whatever. So lying down, we heard a great shout outside. Into the cockpit bolted the skipper, and there, just astern of us, was a big lugger, close-reefed, and seemingly full of men. With splendid ease she was stayed and put close to us to leeward, someone aboard asking if we wanted assistance.

" How much will you put a man aboard for? We want to get our anchor," sang out the skipper.

" Fifteen pounds," came back the reply.

"Go to blazes!" returned the skipper, who was used to these little ways. "I'll give you a pound."

Whereat no more was said, and the lugger plunged on her way, stayed again, and stood in towards Margate.

Before going below the skipper took a look round, and saw that the Roads were filling with craft of all sorts. A steamer or two anchored far out, and nearer inshore some small coasters. Among them dodged some smacks, English and French. The weather was really bad, he thought, but not so bad as to cost fifteen pounds to get out of—at least, not just yet. So he once more pumped the boat out, and then went below to await events and to try to get a smoke.

Knowing the fishermen of these parts pretty well, he felt certain they would come and try to bargain. And he knew well too that if actually driven from his anchor, or in danger of foundering, they would be off at the first signal of distress. So he waited, and chuckled a little as he sat; and though it grew dark, and as the darkness grew the wind increased again, he felt more cheerful than he had done for many hours before.

So he did *not* hoist the riding light this time, for fear they should think it a signal of distress. The light through the cabin window would show them ashore that we were still on top, and would guide them when they came again.

But they came no more that night. The weather grew worse, and the boat rode heavily. Going out to pump again about eight at night, the lights of Margate were still visible, blinking cheerily through the spindrift. The yacht held her position splendidly, and *we knew where we were,* which lifted a great weight off our minds.

Boats are surprising things. Here was this aged packet lifting and plunging and rolling, leaking steadily all over, yet bravely facing the sea, and making fairly good weather of it. Of course she was light, having no heavy weights aboard, with a good bold side to lift her over the swell, and to smash the little cresting tops to spindrift. And that good little anchor was holding away down below like a Briton. But it was a horrid long night. One could doze for a short time, and then awake with a start, notice that the fire wanted a bit of coal, and long for a drink of water (we were quite out of it now). But there was only Nectar, and some Benedictine—neither

of which seemed at all inviting or in the least appropriate to the occasion.

The Poet would rouse now and then, stagger out into the cockpit, and look round in a dazed manner. "Haven't you had enough of this, old chap?" he said once, and only once. I had, but, as our French friends would have said, "Que faire?"

The dawn came, and with it less wind. Once more the yacht was pumped, and the waiting resumed. About eight o'clock the swell seemed less; no doubt the ebb made a little difference. As the skipper was nibbling a bit of biscuit he heard a shout outside, and very near—a great thump, and a hearty British curse! Opening the cabin doors he saw a man in the cockpit, and a big shore boat close by with three men in her pulling away for the shore. The man was rubbing his leg, having fallen over the sail bag, which he had landed on. Looking up he said, "'Ullo, Elbert Strange. Why the 'ell are you stickin' out 'ere?"

"Well, Bowen, I can't get my anchor, and I don't want to lose it. How on earth did you get aboard?"

"Jumped out o' that there shore boat. I knowed the yot, and got 'em to pull me off. It'll cost yer ten bob."

"All right. Come below and have a drink."

"No —— fear; not me. We ain't got no time ter waste. Wot sail d'yer think she'll stand?"

"There's three reefs turned in the mainsail; she'll stand that."

"No fear; we'll goose-wing her, that'll be enough to ratch in with. I knows this old lady; she don't like canvas in a 'ard wind like this."

"All right, any way you like," cheerfully assented the skipper.

So we goose-winged[1] the mainsail by lashings from the leach to the boom, which was just as well, considering the sail was pulling out of the roping at the through-reef[2] cringle. Then we tailed on to the anchor and pulled like Trojans. Bowen was a very powerful man, and I was no weakling for my size, but we couldn't move it.

"Wot's that other bloke doin'?" Can't he pull?" said Bowen at last.

"Well, we'll try him," replied the skipper, doubting if the Poet would be of much use.

And the three of us tailed on, got her pinned with the chain nearly up and down,

For footnotes, see page 238.

and waited a bit, all of us wet through to the waist. Presently a bigger sea came along, the boat's bows went under, but rose again with the anchor free.

" Got inter a crack in the chalk," grunted Bowen as he pulled the chain in, all as bright as silver. " 'Ard up!" he sung out as he hoisted the foresail, and in twenty

water cask and brew a big jorum of tea. Then we squared up the boatmen who had put him off, and rewarded him. We were storm-stayed here four days, during which time Bowen repaired our mainsail and made good other damage to sails and gear. When Bowen's barge had discharged her cargo of bricks we left together for Graves-

Our last sail in Quest.

minutes we were inside the pier with her forefoot on the mud.

Bowen was a waterman of my acquaintance, a hard-bitten man, who had shipped as a temporary hand on a barge which was now in Margate. We had befriended him and his mate last June at Ramsgate, where he was looking out for a job in a waterman's skiff, and had thus cast our bread on the waters in a most profitable manner, seldom realised in this wicked world.

We lay alongside his barge, and the first thing we did was to fill our kettle from his

end with a light south-east breeze, and that evening locked into the Canal Basin, and our long voyage was over.

Quest sailed very little after this. We sold her to a beginner who wanted a very cheap boat, and the very first time he took her for a sail he put her ashore on the chalk embankment in Sea Reach, where she bilged and sank. I warned him that she was very slow in stays and wanted plenty of room to come round in when he bought her. But I suppose he must have forgotten.

If anyone will take the trouble to look up the weather records about March 15, 1882, he will learn from them that this breeze was a real gale, in which much damage was done and many lives lost. The trees round our house in Kent were uprooted, and we were given up as lost when no news arrived from us after leaving Broadstairs. It was not the sort of experience one would have sought voluntarily, but it did me good; and as for the Poet, I think he has never since set foot on any boat. We may have been rash and foolhardy to have come across the Channel at all at that time of year; but there, I feel that I can say with Hilaire Belloc, after his escape from a North Sea gale, told in his splendid book, " Hills and the Sea " : " But which of you who talk so loudly about the island race and the command of the sea have had such a splendid day? I say to you all, it does not make one boastful, but fills one with humility and right vision. You will talk less and think more. Read less, good people, and sail more, and, above all, leave us in peace."

---

1 *goose-winged* – An old meaning of the phrase not now commonly in use. Dixon Kemp (11th Edition) describes the principle: 'Goose Wings – The lower part or the clews of sails when the upper part is furled or brailed up; used for scudding in heavy weather.'

2 *through-reef* is probably a mis-print for *third-reef*.

CHAPTER 6

A Cruise on the Elbe and Baltic

The Yachting Monthly, July and August 1908

Geo. Holmes   *Humber Yawl Club Yearbook, 1909*

A CRUISE ON THE ELBE AND BALTIC

Albert Strange joined forces with his great friend George Holmes for this cruise aboard the celebrated canoe yawl *Eel*, which Holmes had designed for his own use in the previous year. At the age of 22, Holmes was a founder member of the Humber Yawl Club in 1883 and Strange's junior by six years. The two men had quickly become firm friends when Strange joined the HYC in 1891, united by their common interests in painting, cruising, and cruising-boat design.

The account which follows first appeared in the July and August 1908 numbers of *The Yachting Monthly*. As with 'A Winter's Tale', the cruise described took place some time before, but in this instance only 11 years separate the account from the events in 1897. Strange had previously written a short account of the cruise for the 1898 edition of the *Yearbook* of The Humber Yawl Club. Accompanied by a page of sketches, this brief account was chiefly aimed at providing details of the costs and logistics of a foreign cruise in this particular area for the benefit of other members of the Club.

Such cruises in foreign waters were not new to members of the HYC. In the 1933 (50th anniversary) edition of the *Yearbook*, Holmes wrote

> We used to cruise foreign in those days, as it was quite a simple matter to put a 14ft canoe-yawl aboard a passenger liner or a tramp steamer and take her to the foreign port from which you wanted to start and if time was running short, there was no great difficulty in putting the boat aboard a local steamer and doing part of the cruise that way. The Cassy [13ft x 3ft 4in] went to Gothenburg and then up past Trollhatten to Wenersborg by local steamer, rowing and sailing across Sweden from there, and again by steamer from Mem to Stockholm from which place she returned by train in an ordinary guard's van and from Gothenburg by steamer again.

The boats gradually got bigger, perhaps reflecting the increasing years of their owners and their desire for greater comfort. Strange's *Cherub II* of 1893 had a lifting cabin on a 20ft hull of just over one ton displacement. Holmes' *Eel* of 1896 was 21ft 7in with a fixed cabin and displaced about two tons. This was about as big a boat as could be shipped economically aboard a steamer, but, of course, transport by standard railway wagon was no longer possible.

Holmes had previously cruised in the Danish part of the Baltic in 1894, in the two 18ft canoe yawls, *Daisy* and *Kittiwake*.

After the Elbe and Baltic cruise here related by Strange, Holmes returned to Hamburg with *Eel* in 1908. This time, the Elbe-Trave canal from Lauenburg to Lübeck (which he and Strange had discovered to be more a rumour than a reality in 1897), was complete. He used it to enter the Baltic at Travemunde and cruised some 200 miles eastward as far as Stettin.

Historically it is interesting to note how a certain patient and amused tolerance of German authority and officialdom runs through Strange's account. This is common to the writing of several English yachtsmen at this time and, indeed, between the two World Wars.

Away from the water, it finds its satirical climax in the comic novel of Strange's contemporary, Jerome K Jerome, whose three heroes of *Three Men In A Boat* take themselves on a cycling holiday in Germany in a sequel, *Three Men On the Bummel*, published in 1900. The anarchic progress of the three unruly companions is totally at odds with the portrayed German love of officialdom and order. By contrast, Erskine Childers, who had made several cruises among the Friesian Islands, the German estuaries, and the Baltic in the years immediately preceeding 1900, was motivated by his strong interests in politics and military strategy to write his famous propagandist novel, *The Riddle Of The Sands*, in 1903. He uses this powerful novel to point out the dangers of a German invasion of Britain's undefended East Coast – for Childers, Kaiser Wilhelm II's aggressive imperialist intentions were plain and the time for tolerance and satire over.

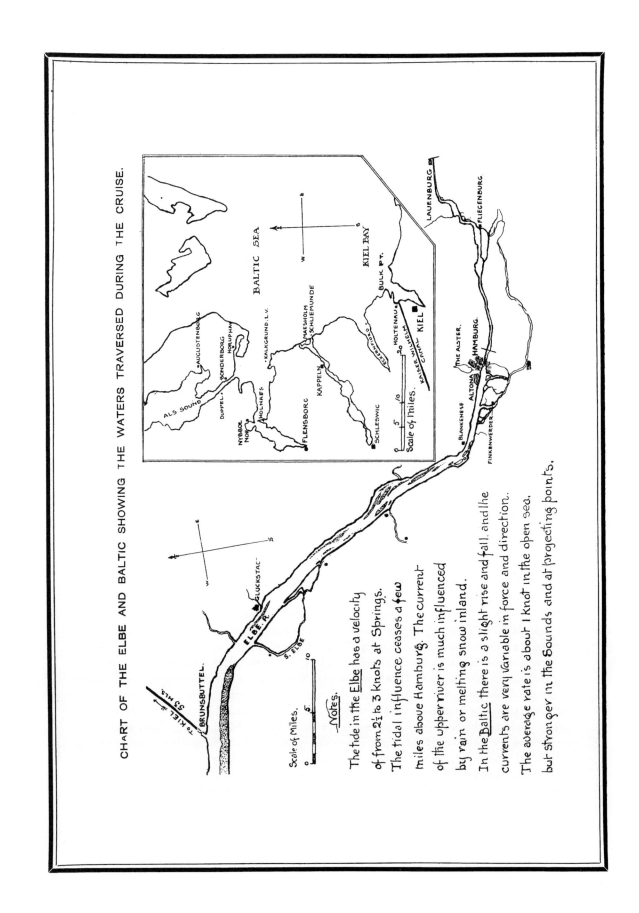

CHART OF THE ELBE AND BALTIC SHOWING THE WATERS TRAVERSED DURING THE CRUISE.

Notes.

The tide in the Elbe has a velocity of from 2½ to 3 knots at Springs. The tidal influence ceases a few miles above Hamburg. The current of the upper river is much influenced by rain or melting snow inland.

In the Baltic there is a slight rise and fall, and the currents are very variable in force and direction. The average rate is about 1 knot in the open sea, but stronger in the Sounds and at projecting points.

BALTIC SEA

ALS SOUND
AUGUSTENBORG
SONDERBORG
DUPPEL
HORUPHAV
NYBBOL NOR
AHOLNAES
FLENSBORG
KALKGRUND L.V.
KAPPELN
MAESHOLM
SCHLIEMUNDE
SCHLESWIC
ECKERNFORD
HOLTENAU
KIEL
KAISER WILHELM CANAL
BULK PT.
KIEL BAY

Scale of Miles.

LAUENBURG
FLIEGENBURG.
THE ALSTER.
HAMBURG.
ALTONA
BLANKENESE
FINKENWERDER
GLUCKSTAD
ELBE R.
S. ELBE
BRUNSBUTTEL.
To KIEL 53 MLS.

Scale of Miles.

A CRUISE ON THE ELBE AND BALTIC.

BY

ALBERT STRANGE.

IN the midst of a galaxy of shining virtues the one outstanding merit of a Humber yawl of the larger class is the ease with which she can "go foreign." The Eel, though weighing two tons when fully loaded with her cruising kit, and yet only drawing 2ft. of water in this trim, is easily hoisted aboard and stowed on deck or in the hold of an ordinary coasting steamer, and thus can make the passage perilous of the North Sea in a little over twenty-four hours, and at the end of that time find herself and her crew afloat in foreign waters in spite of whatever the elements may have to say to the contrary. This feature is of enormous value to men who have all the desire and ability but none of the time necessary to cruise foreign if the oversea passage has to be made on his vessel's own bottom.

And so, after leaving Hull at midnight on Aug. 7, we were all ready to cruise anywhere about the Elbe at 1 p.m. on Aug. 9, after having bidden farewell to the captain and crew of the S.S. Esperanza, who had looked after us well on the voyage from Hull to Hamburg.

My old friend George Holmes, owner and master of Eel, had offered me a berth as "deckie" for this cruise. We had a good fit out of charts and every necessary, including passports duly endorsed by the Consular representative of the German Empire at Hull, and we promised ourselves a good time for the next three weeks of our lives.

We were old companions in cruising. On waters fresh and salt, English and foreign, in fair weather and foul, our "two hearts had beat as one," as the poet puts it, notwithstanding my sad reputation

On the Alster.

as chief rainmaker and friend and companion of thunderstorms, gales, and tempests.

It was, therefore, thoroughly in keeping with my bad character that we navigated the boat under a steady downpour of rain through narrow and tortuous canals, under bridges and tunnels, from the Sandthal Quay, on the Elbe, to the beautiful Alster, a lake in the heart of the most aristocratic suburb of Hamburg. Somewhere on the way we picked up a member of the Hamburg Yacht Club, who, amongst other qualifications, spoke good English and was a connoisseur of Scotch whisky. This kind soul towed us to a berth opposite a floating club-house, where we anchored for the night, and made us free of the aforesaid club for the time being. He was also good enough to warn us against drinking any water whatsoever during our

suggested by the far-off murmur of traffic, the merry voices ashore, and the crowded little passenger steamers for ever crossing and recrossing the lake. Only the cruiser in small craft can enjoy a contrast such as this, and after a long draught of pure enjoyment we, in our little floating island of English oak and larch, turned in at last and slept the sweet sleep of the labouring man, after our exertions in getting the boat up from the river to where we lay.

Next morning we took steamer to the pier under the great Town Hall, and lunched at the Hamburg Yacht Club. This club has its rooms in subterranean chambers beneath the Town Hall. The dining-room is very mediæval in appearance, and the walls are hung with half models and draughts of its more celebrated yachts. We were a merry

On the Upper Elbe, near Lauenburg.

stay in his country. This kind advice was acted on by at least one member of the Eel's crew. It was no hardship; there is really no necessity to drink water in Germany, as there are innumerable kinds of the most excellent beer in the world. This beer is of all shades of colour and flavour; there is no occasion to grow tired of it, a change of brew being sufficient to ward off all feelings of monotony.

What a real and splendidly thorough change it is for a tired man to shift his being to somewhere new and foreign! Here we lay in our snug little ship on this still and shining lake, whose waters stretched away into the darkness of the windless night. Around us floated soft strains of beautiful music from an invisible orchestra hidden in the tree-covered slopes of a bay a little way off. The rain had ceased, and innumerable lights twinkling far away in the dim obscurity of the night jewelled the dark waters of the lake with their radiance. The riches and splendour of this great city were

company—nearly all young, and all enthusiastic. But one felt the curse of the Tower of Babel and its results, as few spoke English well enough to be able to converse with me, and I possessed no knowledge at all of German.

But Holmes, with much exertion and mental strain, conversed with many, whilst his "deckie" found plenty to see and think about in the models on the walls. The food was weird in the extreme, but, at the same time, highly interesting and quite satisfying.

Then we went and saw pictures in a fine gallery on the borders of the lake, and afterwards received visitors, one of whom was a beautiful maiden, the very personification of a youthful Germania. She was delightful to look upon, spoke shyly a little English, and steered the Eel like an angel. Her summers in this world numbered only thirteen, but she handled the tiller like a veteran, admired our little cabin, and enjoyed our English tea and biscuits.

Nightfall on the Lower Elbe.

245

At nightfall more delicious music, as we sailed the boat in and out of the little bays round the lake, but we felt, as we turned in that night, that a continuance of these delights would be damaging to our moral fibre as hardy cruisers. Into the unknown we determined to push on the morrow, which dawned grey and windy and found us pitching a little on the tiny waves of the lake. During a turn up and down to take a final look at it all we were nearly run down by one of the little steamers. This was from no fault of ours, as we were hove-to pulling a reef down. Although the " deckie " knew no German, he took the opportunity of letting off a fine amount of language of a very loud and Saxon type at the extremely agitated green-uniformed commander of the steam-packet. He was ably seconded by the skipper of the Eel in German, which sounded even more damnatory than Saxon, and then we departed on our way through the canals and the tunnels leading down to the Elbe, both fully in agreement as to the demoralising effect of steam on the German mariner.

Arrived at the great waterway we found a strong westerly breeze coming down in puffs between the high buildings and shipping on each bank. Holmes, having in his mind the idea of exploring a canal

Lauenburg.

which he had heard joined the Elbe at Lauenburg, seized the opportunity, and we ran eastward on the last of the flood until we were near a railway bridge spanning the river on the outskirts of the town. Here we were pursued and overhauled by an official-looking steam-launch which made fast alongside, and then we began to dimly realise what it means to the private individual to have to live in a Custom House-ridden land. Explanations were made and our papers produced, and then we were taken to one of the numerous big Custom Houses which form an impenetrable ring-fence round Hamburg.

I felt a little glad now that I knew no German, and was left on board, guarded by an underling official; whilst poor Holmes was led off by the superior officer to the Hall of Torture. He was

away about half an hour, during which time the underling smoked a pipe and looked at me. I also smoked a pipe and looked at him. When my skipper returned, a rather perfunctory search was made, and then we were released and went on our way, not altogether rejoicing in a suspicion that every inch of our future progress through this happy land would be watched and recorded.

Soon the wind and tide failed us, and a drizzle of rain came on. We let go anchor opposite a large board on the bank which had something that looked like very bad language painted on it. Holmes explained that it meant "anchorage forbidden," and we wondered if our disobedience of its imperial command would result in exile or execution, or only the local substitute for " 5s. and costs." It did neither, but at teatime, just after we had lifted anchor and worked our way a wee bit above the alarming notice, a semi-official-looking person drifted down alongside in a punt and, hanging on to our rail, commenced a conversation in quite good English. My heart smote me, and my prophetic vision revealed a spy; but Holmes, with his great command of *sang-froid*, and aided also, I firmly believe, by the gorgeous H.Y.C. badge in his cap, impressed this visitor with a proper sense of his position. The " deckie " had no badge, and wore a shockingly old Norfolk jacket. The spy, therefore, deemed him a person of no importance, and one unworthy of even the consideration of suspicion. The lowly state hath its compensations after all, and so he went below and pretended to repose, but listened to the conversation with an unclosed ear.

It did not amount to very much, and presently the spy rowed away, not having been invited on board and having nothing fresh to record in his notebook.

We spent the night there at anchor, and early next morning were under way, with a light fair breeze which enabled us slowly to overcome the strong current. The river here seemed to have a fair depth of water, but was rather in flood, and in times of drought the navigation would be intricate, as there were numerous small buoys which we supposed marked shoal places. The banks were low and sedgy and lined with farmsteads and orchards. On some of the farmhouses were storks' nests, but we could see no babies in them, and here and there at intervals a rural "pub" sported national colours on high flagstaffs. At times the river widened out into great lakelike spaces, and the shores retreated into the far distance. The beautiful green-tinted water bore many market

afternoon, when a thunderstorm arrived with its own vicious squall (just to keep the "deckie" in countenance), before which we ran a mile or two. Towards evening we decided to bring up opposite a village perched on a tree-covered hill, and the "deckie" was sent on shore to purchase bread and milk, post letters, and endeavour to improve his knowledge of German, whilst the skipper proceeded to sketch a very gorgeous thing in sunsets that Nature was just then performing.

The name of the village turned out to be Fliegenburg, which name was amply justified if it

Sunset on the Upper Elbe.

boats sailing to or from Hamburg, and long strings of Rhine barges were slowly ascending the stream, towed behind rather powerful-looking steamers, the black smoke from whose funnels formed fantastic wreaths against the pure tints of the morning sky.

Altogether it was a fascinating day of quite new sights and sounds, and with a freshening breeze we ploughed along over the strong current (and once over a submerged stone dyke that gave us a start as it banged the plate up in its case) in delightful ease. For hours we went on, steadily covering the ground, passing village after village until late

implied any connection with flies. My only other remaining impressions of the place are that the female population was large, inquisitive, and extremely plain, and that the juvenile part of it threw stones with quite remarkable inaccuracy.

Also, that when one's first bashfulness in a strange place has worn off, signs are a pretty good substitute for spoken language, especially with the gentler sex. The letters were posted, the stores bought, and the "deckie" returned to the cosy interior of the Eel in safety.

On the Elbe, above Fliegenburg.

We slept with one eye open that night, as Holmes had drawn the boat into a little creek, and we had a suspicion that the river was falling. There was plenty of water when we turned in, but the faithful "deckie" arose somewhere about midnight and carefully sounded, finding about 6ft., which was more than ample for us. Still, one has to exercise caution in navigating strange waters, and the charms of Fliegenburg were not sufficient to compensate for a compulsory stay there, and the resources of the village hardly seemed equal to getting us out if a rapid fall in the river left us aground. And so we stole away early in the beauty of the next morning, and ran on, still with a fresh favourable breeze, up this most beautiful river, until, far, far off, we saw a filmy thread of a bridge crossing the horizon, and knew that we were nearing Lauenburg.

To the true cruiser the discovery of a new place has a most exhilarating charm. We had never in our lives heard of Lauenburg until we looked at the map of the Elbe, and as we reached up close to the place its old-world beauty was an exceeding great reward. It seemed indeed to be

A rose-red city, half as old as Time,

with its leaning, mysterious, half-timbered houses peering down into the blue-green waters of the Elbe. It bore hardly a touch of the great modern expansion of change and prosperity that is so evident throughout Germany. Close by the great railway bridge that spans the river here it is true there were evidences of progress, for where we at last moored they were building big iron barges, and a little ferry motor-boat throbbed and stunk its way across the stream, with a cargo of weather-stained peasants from the southern bank.

But when we had supped and wandered ashore to look more closely at the place, we seemed to have stepped straight into the heart of the sixteenth century. The narrow cobble-paved main street that ran parallel to the river was full of old houses that leaned their upper stories towards each other in a most confidential way; and the warm, almost solid, smell of generations of bakings and brewings, so characteristic of old cities, enveloped us completely.

We wandered along, astonishing the inhabitants by our quite outspoken expressions of admiration, until, at the very end of a beautiful progress, we came upon the "Fischer's Inn," where the "deckie," completely overcome by his feelings, demanded to be led inside to recover from his excess of emotion.

Humbly following the great Dr. Johnson, I have always felt a sincere affection for inns—when they are really genuine places of repose and refreshment, and not like so many of the modern substitutes of to-day from which one's only desire is to flee in haste. Here was my ideal inn—oak-timbered, oak-panelled, oak-floored. Its very walls seemed rich with the jests of long-departed burgomasters, and full of the gracious spaciousness that only the old builders knew how to give. There was, moreover, a quite nice landlord's daughter who brought my beer, and Holmes his coffee. Nothing was lacking. The place was full of a holy peace and a splendid, large odour of sausage and smoke.

In the great joy of the moment I told my skipper that he could take the Eel back home himself if he liked. For my part, I intended to stay here for ever. I had at last found the place of my dreams, and nothing would induce me to leave it. No more "England, home, and duty" for me! There was Teutonic blood in my veins—quite a large quantity of it, and I had found the city of my affinities where I intended to stay for the remainder of my existence.

But Holmes merely suggested that it was getting late, and that I might at least spend one more night on board the Eel. He also added, rather

LAUENBURG. Elbe.

unfeelingly, that I should probably feel better in the morning. Such is sympathy!

As we returned to the boat we walked on opposite sides of the street and sought our respective couches in meditative silence.

We were both very busy next day exploring and sketching the remoter crannies of the place. The inn was as desirable in my eyes by daylight as it had seemed to be in the evening glow. Right across the street there stretched from its grey old face a most gorgeous wrought-iron sign, crimson and orange with three hundred years of rust. I really should have loved to have bought or stolen it, but Holmes thought it would not stow away very comfortably on the Eel, and on thinking it over I do not believe it would, as it was about 15ft. long; but it was a beauty, and to allay my disappointment my skipper presented me with a brown earthenware jug which still abides with me as a memento of Lauenburg.

The canal which we expected to find did not materialise. It was a myth so far as Lauenburg was concerned, though I believe they were thinking of commencing it at Lubeck. Our idea had been to get through to the Baltic by its means, but we were now forced to go back through Hamburg and down the lower Elbe to Brunsputtel,[1] where the great Kiel canal commences.

It was with great reluctance that the "deckie" permitted himself to be torn from Lauenburg, though Holmes did it with the greatest tact and kindness, taking the boat to the south side of the river, whence the disconsolate one could get a fresh view of the town as the moon rose. After some sketching, the poor "deckie" made a final visit to bid his adieux to the much-loved inn and its occupants, going across in the motor-boat and returning, in tears, to find the skipper fast asleep. We sailed down to Hamburg in much quicker time than we sailed up, having a strong current with us and a light reaching wind, and arrived about low water, rode out the flood, and then went on, after another compulsory visit to the Custom House at the western end of the town, which was again quite a long affair; then sailed away on the ebb until darkness found us abreast of Altona, where we anchored close in with a group of trading vessels under a glorious calm, moonlit sky.

Germans of the humbler class seem to have a passion for short steamboat voyages to the various Biergarten which are so numerous in and around Hamburg. On these voyages they are accompanied

Fliegenburg.

1 Brunsbüttel

by their families, and there is always a large brass band which plays nothing but polkas in a very loud key. They appear to enjoy themselves immensely, but to us on the Eel they were a perfect nuisance. When six of them pass you at one time, each foaming along at about nine knots and raising a large bow and quarter wave, the jumble of sea they leave behind is something awful, and not only shakes all the wind out of the sails, but chucks everything about down the cabin and in the lockers. These thoughtless proceedings annoyed the skipper very much, and he grew very angry at the shaking-up his dear boat received.

We found a strong head wind blowing next morning, and turned away down over the weak ebb very merrily. We passed many small fishing craft rigged with balance lug and a jib. They were a good deal bigger than the Eel, seeming about 28ft. or 30ft. long, with leeboards and lots of beam. But they got to windward no faster or drier than we did under our snug sail. By-and-by the tide was done and we had to anchor under the lee of a sedgy island to ride out the flood. Rather before high water we got under way again, to turn over the last of the tide. By doing so we sadly deceived a small coaster, who, seeing Eel doing good business to windward, promptly got his anchor and endeavoured to do likewise. But not being quite fashioned for that kind of work he dropped steadily to leeward, and finally had to anchor again.

Towards evening we were off the mouth of the South Elbe, but could not get far in owing to the boiling ebb coming out, the wind having dropped. So we let go anchor, again in company with a number of coasters. The tide was so strong that it ran in large overfalls, which rendered cooking quite difficult and annoyed the skipper very much. I tried to comfort him with a quotation from "Erewhon" to the effect that "exploring is delightful to look forward to and back upon, but it is not comfortable at the time, unless it be of such an easy nature as not to deserve the name." But I had no success. Samuel Butler's wisdom was received with snorts of contempt. Some time after dinner the rolling grew less, and presently all was still.

In the grey of early morning we reached across to Brunsputtel and entered the great locks of the canal which was our road to the Baltic. After being duly examined and inspected we were let through, and paid the sum of 10 marks for dues, 1 mark 60pf. for tonnage, and 2 marks for a wonderful book and the services of an English-speaking broker, who piloted Holmes through the Hall of Torture and brought him back intact.

For some reason which we were unable to discover we were not allowed to sail, and were the last of a string of vessels of all descriptions, including a Welsh schooner, that followed the small tug which headed the procession up the grey straight waterway of the great canal.

A Pleasure Party.

A CRUISE TO THE ELBE AND BALTIC.

BY

ALBERT STRANGE.

THE skipper of the Eel is a man deeply learned in his craft of seamanship and full of fore-thought and resource. When he beheld the tow ahead and felt the force of the strong wind astern he bade the "deckie" get the end of the towrope on board and pass the bight only to vessel ahead. "For," said he, "there'll be a mess presently when they have to stop, and we shall see a little fun, but it is best to see it from a safe distance, as we are only a little 'un."

The tug went ahead, and we formed into line astern and proceeded at not more than four miles an hour up the canal. We could have made the run in the day had we been allowed, but the authorities seemed to think that if Eel went as fast as she could she might damage the banks of their beautiful canal or run down an approaching iron-clad. So we had to submit to being dragged instead of bounding along at six and a half knots on our own ; and dragged we were from about noon till somewhere about five o'clock, when all at once without any warning the tug stopped opposite a long pile-lined "lay-by." It was easy to stop the tug, but quite otherwise with its tow, which went on

On the Kaiser Wilhelm Canal.

and got into a most lovely mix-up, some letting go their anchors, others running foul of the anchored ones, and for the space of twenty minutes the air was rent with cries and objurgations. Thanks to the skipper's foresight, we slipped at the first alarm and boxed about under jib and mizen until the tow got sorted out and ranged alongside the piles. Then we ran alongside the British schooner Lizzie, of Port Madoc, feeling quite at home under the folds of the dear red ensign, where we stayed during the night.

The skipper of the Lizzie invited us on board,

In Kiel Bay.

and we played draughts ; in which game the "deckie" got badly worsted, but I think Holmes held his own with the Welshman, who was a kind-hearted man and gave us a piece of beef from his harness-cask, which I boiled, and on which we fed for several days.

At 5 a.m. next morning the passage began again, and we got to the eastern end of the canal about 1 p.m., going straight out and on to Kiel, my dear friend Jupiter Pluvius attending us the while.

We called at the agents' (Schutt & Sieck) for letters and also bought provisions at the same place. Holmes, by now beginning to realise what he had undertaken in having shipped such a "deckie," bought himself an extra patent waterproof coat of a very delicate light-fawn tint, although he had plenty of the ordinary sticky sort on board. It was big and long, and he looked much like a cricket umpire when arrayed in it. He thought it wise to make extra provision against rain on this occasion, for, said he, "If we make this three weeks' cruise together, our ordinary oilies will have been completely washed away before the end."

Kiel is a charming place, ideal

On the Baltic Shore.

from all points as a harbour and most interesting as a town. It was full of soldiers and sailors, and we saw for the first time on the cruise what a tremendous amount of "side" the German officer can put on when he appears in public. It really seems quite natural and in no way assumed, and the civilian population adore it, especially the ladies. We had an indifferent dinner at a big restaurant and returned on board to the sounds of martial music and illumined by blue lights. I was not quite certain whether these were in our honour, but Holmes thought not. We watched some curious naval manœuvres with critical interest during the afternoon, and felt that we had got to a really grand place for sailing.

We were off early next morning with a nice breeze and bright sunshine, bound for Schliefiord [1] As we .left Kiel we saw the ironclads, some of

rather an obsolete type, going out to gunnery practice at big targets moored in the Baltic, but we thought it best not to watch the proceedings. The coast is very picturesque, not grandly so, but of a tender beauty quite distinctive ; the low rolling hills are dotted with farmsteads and beech woods. Yachts and small craft were abundant, as they ought to be in such a paradise for yachting.

As we went on, stretching across the wide mouth of the Eckernfiord, the shores grew lower and were splashed with bright yellow stretches of sand, which in contrast with the blue sparkling sea dazzled the eye. Great towering cumulus clouds filled the sky to the north-west, and these were the heralds of a thunderstorm in which the skipper's new coat was put to the test very thoroughly. It only lasted a short time, and passed away grandly across by Bulk Point, leaving us with a little trickling breeze which just allowed us to reach into the very narrow mouth of the Schliefiord, which has a lighthouse and piers and buoys all complete. Contrary to our expectation, we were permitted to enter unmolested by coastguards or custom-house people, and we proceeded to make the best of our way up the very narrow channel, which was marked by saplings stuck into the mud. Sad to say, the "deckie"—who was taking his trick at the tiller—stuck the ship on shore, because he could not keep his eyes off the beautiful scenery. He got her off without too much exertion, and received the admonitions of his superior officer in respectful silence.

We turned up this lovely fiord with a smart breeze which gave us every opportunity of examining each shore. Such an ideal place for yachting as this is seldom met with by cruisers, but Denmark seems full of them. The fiord is thirty miles long from the mouth to the head, and is of various widths, opening out at times into great lakes called brednings. Yet it is almost deserted, save by a few fishing-boats and coasters. I think we only saw one yacht, moored in a snug bay on the starboard hand. If England only possessed a

1 For footnote, see page 260.

few places like these fiords, what a wealth of boats and yachts would cover their waters ! I thought of our rough North Sea and our tide-vexed Humber, and envied the Germans their vast wealth of sailing-grounds.

In the afternoon light Kappeln looked very pretty sitting on the shore with its back to the sun. We moored alongside a rough jetty, just above a row of bathing-boxes, for if the natives do not sail much they bathe a good deal. My skipper was much shocked when I pointed out to him a group of damsels who, innocent of any bathing costume, were splashing about in the shallows on the other side of the fiord. It still seems to be the

English, the "deckie" and he foregathered and held a communion of soul, if not of intelligible speech.

Next day we sailed down to Maesholm, an island at the mouth of the fiord, inhabited by a race of fishermen who are very different in many ways from the landsmen of these parts. The island is small, and the village consists of a single picturesque street. The craft they use are open spritsail boats, of light draught, sharp stem and stern, not unlike the general run of Baltic fishing-vessels. They would, we thought, have been improved by the addition of a centreboard. After busily sketching the place we left, late in the afternoon, and sailed along with a pretty little off-shore breeze until we

Kappeln, Schleswig Fiord.

custom of the country, and was noticed by the Rev. R. E. Hughes in his account of the cruise of the Pet, written in 1855. "*Honi soit,*" etc. The damsels were not immodest ; they were only natural —and at least 150yds. away. A figure-painter might have made a fine subject of the incident if he could have persuaded them to stay whilst he sketched. Not being a figure-painter I did not try.

In the twilight we were visited by a retired sea-captain, who came on board and talked, and next day showed us the town from the heights of the church-tower, and took us home and gave us glasses of some mysterious sweet wine. He was a dear old soul, and vastly admired the way in which Eel turned up the fiord. Kappeln was a most interesting little town, and possessed a photographer who was a real artist. Though he did not speak

were abreast of the Kalkgrund lightship, when the wind fell light. We kept on across Flensborg Fiord, and in the gathering darkness entered Als Sound and groped our way into a corner of the harbour of Sonderborg, where we spent an uneasy night, the "deckie" succumbing to a mysterious illness, and the Eel bumping first against the piles and then against a fishing-boat. All night long it poured in torrents, but the "deckie" remembers little of what went on outside until he found himself off Duppel Hill next morning, when he and his skipper visited the monument which marks the site of the final sanguinary struggle between Germany and the Danes in 1864. The victors have displayed that best of qualities towards a vanquished foe— generosity, and no difference is made in the honour given to the fallen. "Here lie many brave German and Danish soldiers," says one inscription, and many others mark the burial-places of the combatants in the same simply pathetic manner.

It is a little touching to an Englishman to stand on this great battlefield, remembering all the ties of blood and marriage that bind the Danes to England. No doubt the cold decrees of interest correctly forbade us to move a finger to prevent or moderate the bitter struggle—and it is doubtless true that at least half the inhabitants of Schleswig-Holstein were pro-German—yet no Englishman who knows the Danes can help regretting any event that deprived this splendid little nation of an inch of territory or one particle of influence. No one takes sides now—at least, not

sea like the Baltic there should be such strong currents. But the cause is simple. The prevailing winds are N.W. and S.E., and they heap up the waters in the various fiords, to return to the main sea when the wind shifts or drops. Thus these currents are variable in direction and hard to be foreseen even by the natives. But, as a rule, if you have a head wind you have a lee-going tide, while a fair wind brings a favourable current, a fine nautical illustration of the text, "To him that hath shall be given."

We sketched and loafed until evening, when we turned in early to the sound of pattering raindrops and the whistle of the wind in the rigging, while melancholy bugles blew faintly across the dark

Here lie many brave Germans and Danes.

openly—in this part of Germany, so let us hope that all is now forgiven, if not forgotten, and that the two races are working harmoniously together for their common good.

Als Sound is beautiful—a paradise for yachtsmen and cruisers, but there are drawbacks. One is the enormous quantity of weed that covers the bottom in ordinary depths of water. As we reclined on the hill and ate blackberries we saw that Eel was off on a little cruise of her own, so we hurried down to the dinghy and recaptured the errant craft. On getting the anchor we found it a huge bundle of eelgrass, which we cleared, and then ran back to an anchorage just above the bridge of boats at Sonderborg.

A perpetual current sets through this sound, and it strikes a fresh comer as curious that in a tideless

waters of the sound from the old castle watching high above dimly-lit Sonderborg.

In the morning more rain and plenty of wind, so that the "deckie" was justified in reproving his skipper for tempting providence by buying a pale, unnecessary waterproof. To this cause he attributed the prevailing moisture and the quite superfluous amount of wind, which at last grew so strong that we were reduced to almost bare poles when running into Augustenborg Sound, a place studded with rocks above and below water. Wind and driving rain and scud quite obscured the shores and marks, and we counted ourselves lucky in getting into shelter without hitting something nasty. As usual, no sooner had we let go our anchor than it cleared up, and we walked into Augustenborg itself, a sort of toybox town with precise little mansions and cottages and the usual "schloss." One almost expected to meet the author of "The Prisoner of Zenda" looking for material for a fresh novel of romance. It was just the sort of place where such a *rencontre* would be likely to happen.

After our walk through the town we penetrated the woods with which the place was surrounded and crossed the narrow peninsula until we could look over the waters of the Little Belt. Here was evidence of the tremendous power of the prevailing north-west winds. The trees lining the shore were bent and twisted into the weirdest shapes, and fancy could easily paint the winter struggle between woods and winds as here depicted in their gnarled and tortured trunks.

Augustenborg, charming as it was, did not detain us, and next day we retraced our way through

It seemed a sporting thing to do our best to beat her, and this we did without much difficulty, and we were rather disgusted to see her give up so soon. Presently the skipper, who was looking at the chart, said, "There seems a nice little place in here to leeward; let's go in and have a look at it." So "deckie" started his sheet and we were soon inside. It *was* a nice little place with a charming name—Ekensund—with a village equally nicely named—Nybbol Nor—this last having quite an English sound about it. It was a perfectly land-locked circular bay with a large shoal in the centre. We ran in and brought up in about two fathoms, strolled ashore, and found that the local industry was the making of bricks, and apparently very flourishing, as many craft were loading these necessary items in house-building. It was quite a

Als Sound, passing on our journey another school of native mermaids and also a smart sprit-rigged open boat about our size, after quite a sporting match. We were handicapped by our dinghy, but, after a very close series of boards, finally weathered our friend, and then, the wind hardening, we went away fast.

We passed through the bridge of boats and soon were out in the broad waters of Flensborg Fiord, where we found plenty of wind, and the water began to fly. It was a hard turn to windward with some jump of sea, and at Holnaes we passed, inadvertently, a custom-house, where we should have stopped for the usual examination. It did not occur to us that we should be required to stop, but a sailing craft came off—I suppose in chase of us, though again this we did not know.

picturesque place and generally beautiful in colour. We sketched that afternoon and again next morning, and then decided that we must go on to Flensborg. Just as "deckie" was getting the anchor a boat rowed off to us. In the boat was a green-uniformed gentleman with a sword, who seemed angry and who shouted something in a very peremptory manner and then came alongside. Holmes, as usual, took him in hand, but soon found out that something serious had happened this time, for we were forbidden to leave. The green-uniformed one talked, gesticulated, and perspired freely, and "deckie" prepared to gather up his belongings in a handkerchief and go meekly to the awaiting dungeon, for there seemed no way out but this, and Holmes's German, though quite good for ordinary purposes, was not profound enough to win us through a case so serious as this was turning out. I daresay we should be there now had not an English-speaking chap turned up, who acted as intermediary, and whose words seemed to have some weight with the green man with the sword. Again, Holmes was taken ashore with all our papers, and, as usual, the poor "deckie" was left in charge of the ship, half the population lining the shore and gazing sadly and earnestly at the doomed vessel; for things like these are *very* serious in Germany.

from Kiel to Berlin, from Berlin to London, for it was not until 4 p.m. that the English-speaking one came alongside and said it was all right. *He* had chartered us to Flensborg as a light cargo vessel, and, of course, there was brokerage to pay. We paid, and left unmolested but greatly wondering, not breathing freely until we were well clear of Nybbol Nor and on our way to Flensborg, though we did not get there that night owing to a calm. We brought up in a delightful little bay almost among the reeds and rushes, and went along in the

Running into Augustenborg Fiord.

It seemed hours before my skipper returned, and then he told me that we had committed a great crime in passing Holnaes, but a much greater one in beating the vessel that came after us. We were not to go yet, though I believe he had bribed the English-speaking man to do something for us, as he held out hopes that we might get away with our lives, if not with the Eel. Anyway, they were telegraphing like mad, and until something definite arrived we might as well feed and go on sketching. We thought they must have telegraphed to Kiel,

early morning to get letters and provisions and final absolution for our crime.

Flensborg offers no attractions to the eye, and less than none to the ear and nose. It is a busy, small imitation of Hartlepool (with a far finer situation), for they were building five or six small coasting steamers of about three or four hundred tons burthen, and this means noise; also the harbour seemed to be the outfall of the main sewer.

We received official absolution for our sins, and left gladly. With a splendid reaching breeze we made good time to the hated Holnaes, where, it seems, one does not have to stop when outward bound, and then began a long turn to windward under double reefs down the wider part of Flensborg Fiord, which here makes a complete rectangular turn, bringing our reaching breeze into a dead knock. As soon as we had passed the custom-station, a revenue schooner of at least fifty tons came along after us. We had three-quarters of a mile start, and, though it is almost unbelievable, she did not catch us until we had sailed the ten miles or so from Holnaes to Sonderborg, where she bore up, whilst we turned on and into Horup Hav. We were very pleased with ourselves, and went ashore and had beer at the "Baltic," the hotel whereat Mr. Knight, of "Falcon" fame,[2] got a present of an eel when he thought to receive a lobster !

Our time was drawing to a close, and we made no long stay at Horup Hav, charming little place that it is. We left with the promise of a hard day's work to get to Kiel, thirty miles away, with a very strong scant wind[3] to do it with. We just managed to squeeze our course (with an occasional hitch or two in-shore) to Bulk Point,

Farmhouse at Nybbol Nor.

little village, with a boat-builder's yard, where a trader of the general Elbe pattern was being repaired. These vessels have a curious under-water form, a very wide keel forming a sort of ballast-box, and then above it the ordinary round bottom. We pondered on this odd feature, but came to no definite conclusion thereon, and departed on our way towards Hamburg with a fair wind and tide, reaching a typical riverside village on one of the many branches of the Elbe—which here cuts the country into numerous islands large and small —late in the afternoon. The name of this village was Finkelwarder.[5] Here many small craft were being built, and the local fire-brigade busily engaged in pumping out the lower story of a house which had been filled by the high tide—a business that must often happen here. About dusk an official of some species rowed off and commanded us to put up our riding-light. As we did not immediately proceed to obey he stood on the bank and shouted at intervals until we had complied. Very kind, thoughtful people these, who take great care that the stranger shall come to no harm within their gates.

On the next day we sorrowfully made our way back to the main stream, and took the Eel alongside the steamer ready to be hoisted aboard. The "deckie," having a horrid suspicion that piles of letters awaited him somewhere in Hamburg, spent a weary afternoon in hunting for the General Post-Office, and visited almost every post-office in Hamburg before he found it. It is really extraordinary how very few persons seem to be able to understand either English or French in Germany. The afternoon was very hot and uncomfortable, and people seemed rather cross when stopped by a stranger and addressed with much politeness in a foreign tongue. The "deckie," who had been brought up in the profound conviction that every little German boy and girl was taught English thoroughly, returned to the steamer bereft of some cherished illusions on the subject of education abroad.

where we found a strong lee-going tide coming out of Kiel Bay, and we were a long time hammering to windward over a nasty short sea. We weathered it at last, and then it was late evening and the wind fell. So it was long after dark when we at last came into the glare of the lights at the Holtenan[4] entrance to the canal, pushed in, had our papers examined, and then found ourselves in a tiny out-of-the-way corner in the darkness, and turned in to our virtuous beds, conscious of having performed an excellent day's work, and more than ever convinced that the Eel is the beau-ideal kind of boat to push a hard passage in.

The passage through the canal to Brunsbüttel was without incident, and we hastened across to the South Elbe, but got farther in this time, opposite the village of Wischaven, clear of the overfalls caused by the ebb at the mouth. This was a quiet

For footnotes, see page 260.

In the evening many emigrants came on board, also many officials, who inspected everybody and everything, and at midnight the Esperanza was on her way to Hull. Somewhere in the early morning we realised that a gale was blowing and the steamer plunging heavily. Wreck and confusion reigned in the saloon, and, worst of all, no breakfast was forthcoming! After waiting hours we struggled along to the galley for'ard, walking over the prostrate bodies of the unfortunate emigrants, who were lying everywhere about the alleyways. We got some coffee and cold pie in the galley and then pumped out the Eel, which was half full of water, and at times almost washed off the deck. This sort of thing went on for about thirty hours, the only amusement being to watch the engines rocking on their bed and to measure the angle of the rolls the ship made. We also visited the mates in their cabins and listened to their remarks on the behaviour of their vessel, remarks not at all complimentary.

At last we crawled into the Humber, some twelve hours late, and, going full speed ahead, just saved our tide into the docks, and our cruise, which had been so intensely interesting and so full of enjoyment, was ended. It would be difficult to plan a better round trip or one fuller of variety. The navigation of the Elbe is easy to cruisers used to tidal river work, that of the Baltic easier still, and to any who may feel inclined to follow in our wake I would only add, "Polish up your German before starting, and don't sail past a custom-house too often without heaving-to." All the rest is simple; the German yachtsman is a thoroughly good sort, so is his Danish cousin, and it will be his own fault if the cruiser does not have at least as good a time as the crew of the Eel enjoyed.

-A. Strange- -Wischaven-

1 *Schliefiord.* The fiord at the head of which lies Schleswig.
2 The reference is to E F Knight's book, *Falcon on the Baltic. Falcon* was a converted ship's lifeboat. In her Knight sailed from the Thames to the Baltic in 1887.
3 A scant wind is from a direction which barely allows the course to be laid close-hauled, without beating.
4 Holtenau.
5 Finkelwarder is spelt Finkenwerder on the map.

APPENDIX 1 ——

Bibliography

Key to abbreviations

AS   Albert Strange

HYC   Humber Yawl Club

JL   *Albert Strange, Yacht Designer and Artist, 1855-1917.*
            John Leather, Pentland Press Ltd., Scotland, 1990.

★Mystic   Mystic Seaport Museum, CT, USA.

RCC   Royal Cruising Club

YM   *Yachting Monthly* magazine

YW   *Yachting World* magazine

★ Many of Albert Strange's original design drawings are in the collection at Mystic Seaport Museum, Mystic, CT, 06355-0990, U.S.A.
Copies may be purchased quoting the reference numbers given in the bibliography.

Where a book is listed in the bibliography the title is given first, followed by the author, country or town of publication, the publishers and, finally, the date of the relevant edition.

Yachts mentioned in Chapter One - An Introduction

BIRDIE later named *Dorcis*. AS Design No. 39./1898.
LOA: 29ft 4in. LWL: 23ft 9in. Beam: 7ft 5in. Draught: 3ft 10in. Built 1899.
HYC Year Book 1900 – Details by Strange.
YM Aug. 1941 – Descriptive article in 'The Other Man's Boat' series with Lines, Sail Plan, and Accommodation.
JL page 49 – Lines and Sail Plan.
Mystic 1.727 – Lines only.

CHARM
LOA: 33ft. LWL: 25ft. Beam: 7ft 7in. Draught: 5ft. An enlarged version of *Venture*, see below.
YM Oct. 1923 – Article describing cruise in *Charm*, Lowestoft to Falmouth and return.

CHARM II
LOA: 40ft. LWL: 30ft. Beam: 9ft 2in. Draught: 6ft (approximate measurements), built 1925.
YM Nov. 1957 – Account of a cruise from Burnham-on-Crouch to the Baltic and return.
Believed to be another enlarged version of *Venture*, see below. She made a double Atlantic crossing in the early 1980s when owned by Dr. Stephen Rogers.

CHERUB

LOA: 21ft. Beam: 7ft 3in. Draught: 3ft. Thought to be the first yacht designed by Strange.
Built 1887/8. Rigged as a gaff cutter. Sold 1896, renamed *Seabird*, counter added.
The Field magazine 28th Feb. 1891 – Letter from Strange commenting on the advantages of
the fore-and-aft centreboards.
JL page 17 – Photo of half-model.

DAUNTLESS

LOA: 18ft. Beam: 6ft 6in. Draught: 2ft. Approximate measurements. Believed to be a converted
Thames peter boat.
Model Yachtsman and Canoeist, April 1893 – Account by Strange of how *Dauntless* came into his
possession.
YM Mar. 1919 – Article on Peter Boats by W Edward Wigfull.
JL pages 14 and 173 – Lines, Sail Plan and further details.

HAWKMOTH   AS Design No. 85./1908.

LOA: 31ft. 9in. LWL: 22ft 3in. Beam: 7ft 7in. Draught: 4ft 7in. Displacement: 4.7 Tons.
Sail Area: 535sq.ft. Yawl rig. Built 1908 by Dickie at Tarbert.
YM Jan. 1911 – Lines, Accommodation/Construction, Sail Plan.
JL page 55f. – Drawings as above.
Mystic 1.679 – Lines, Accommodation/Construction, Sail Plan.

MIST   AS Design No. 78./1906.

YM Mar. 1916 – Descriptive article by her owner. Includes Lines and
Accommodation/Construction.
JL page 79 – Lines, Accommodation/Construction.
Mystic 1.712 – Lines, Sail Plan.

NORMA/SHULAH

See details on page 267.

QUEST

LOA: 36ft. Beam: 10ft 8in (approximate measurements). Designed Hamilton, built by R Cain,
Port St. Mary, IOM, 1902. Sail Area: 1,050sq.ft. Yawl rig.
YM June 1910 and JL page 105f. – Account of 'A Sketching Cruise on the Irish Coast' made
by Strange and owner C W Adderton.

SHEILA   AS Design No. 70./1903.

YM Jan., Feb. 1907. June/July 1908. July/August 1909 – Various accounts of cruises on the
West Coast of Scotland made aboard *Sheila* by Robert Groves, her first owner. Groves was a
noted marine artist and these articles are illustrated with his sketches and anchorage plans.
YW Feb. 1982 – Account by her present owner, Mike Burn, of her complete restoration,
together with comments on various designs of Albert Strange and George Holmes.
Traditional Sail Review Nos. 4 and 5, 1982 and *WoodenBoat* May/June, 1985 contain similar
accounts.
JL pages 74 and 78 – Lines and details of building cost.
Mystic 1.752 – Lines, Accommodation/Construction.

SHEILA II   AS Design No. 117./1910.
LOA: 31ft 7in. LWL: 24ft. Beam: 8ft 6in. Draught: 4ft 11in. Displacement: 6.1 Tons.
Sail Area: 545sq. ft. Yawl rig. Built 1911 by Dickie at Tarbert.
HYC Year Book 1911 and JL page 82 – Comparison by Strange of *Sheila II* and *Cherub III*.
YM Oct. 1912 and JL page 67f. – Lines, Accommodation/Construction, Sail Plan.
YM May 1919 – Cruise made by her first owner, Robert Groves.
Thoughts on Yachts and Yachting. Uffa Fox. London; Peter Davies Ltd., 1938 – Lines etc., comments
and sketch by Robert Groves of *Sheila II* anchored at Lunga, Treshnish Isles, Scotland.
Sheila in the Wind: A Story of a Lone Voyage. Adrian Hayter. London; Hodder and Stoughton,
1959. Published in the USA under the title, *The Lone Voyage*. Describes a single-handed passage
in *Sheila II* from UK to New Zealand via Mediterranean, Red Sea, and Indian Ocean.
Mystic 1.722 – Lines, Accommodation, Sail Plan.

THERESA II   AS Design No. 139./1913.
Built 1914 by Bundock Bros. at Leigh-on-Sea.
YM May 1914 and JL page 129 – Lines, Accommodation/Construction, Sail Plan.
Mystic 1.685 – Lines, Accommodation/Construction, Sail Plan.

VENTURE   AS Design No. 156./1917.
LOA: 29ft 6in. LWL: 22ft. Beam: 6ft 8in. Draught: 4ft 4in. Displacement: 4.3 Tons.
Sail Area: 410sq. ft. Yawl rig. Built 1920 by A Wooden at Oulton Broad.
YM Jan. 1917 – Long article by T Harrison Butler entitled 'The Week-End Single-Handed
Cruiser'. Includes H J Suffling's original design for *Venture,* later refined by Strange.
YM Dec. 1917 and JL page 142 – Final version of Lines and Sail Plan.
The Single-Handed Yachtsman. Francis B Cooke. London; Edward Arnold & Co. 1946. Fold-out
Lines and Accommodation Plan, details and photograph of new bermudan rig designed by
Walter Easton.
Mystic 1.682 – Lines, Gaff Sail Plan.

WREN
LOA: 15ft 6in. LWL: 15ft. Beam: 5ft 6in. Draught: 1ft 3in/2ft 6in. Displacement: 11¾ cwt.
Sail Area: 116sq.ft. Yawl rig. Built 1889 by Jas Frank at Scarborough.
The Field magazine 21st Dec. 1889 – Detailed letter by Strange.
YM Nov. 1918 – Lines, Sail Plan and brief description.

AS DESIGN No. 4./1891
LOA: 18ft. LWL: 16ft. Beam: 5ft. 8in. Draft: 1ft. 8in./2ft. 8in. Displacement: Approx. 1 Ton.
Sail Area: 169sq.ft. Yawl rig. Two examples are thought to have been built, one in 1894 by
R Kemp at Oulton Broad named *Ethel* later changed to *Curlew*.
The Yachtsman magazine 14th Jan. 1892 – Details included in a reprint of a lecture by Strange to
members of the Royal Yorkshire Y.C. on 'Single-Handed Cruisers and Cruising'.
JL page 97 – Lines, Sail Plan.
Mystic 1.733 – Lines, with alternative scales for 18ft. and 16ft. LWL. Sail Plan.

QUOTATION ON THE YAWL RIG.
Sailing Yacht Design. Douglas Phillips-Birt. London: Robert Ross, 1951.

OTHER REFERENCES
Reiach's visit to Tarbert, Loch Fyne – YM Nov. 1910

Chapter Two - The Design and Construction of Small Cruising Yachts

Introductory Article, published in YM, October 1917. Articles I to IX published in consecutive issues, August 1914 to April 1915.

Chapter Three - The Designs

MONA
HYC Year Book 1904 – Lines, Sail Plan, first owner's comments.
YM Dec. 1947 – Comments by later owner.

TAVIE II
The Field magazine, London; 12th Dec., 1896 – Lines, Sail Plan, first owner's comments.
YM Jan. 1913 – Comments on centreboard.

WENDA
The Sailing Boat. D C Folkard. London; Edward Stanford, 1901 and 1906 editions – Lines, Sail Plan, Comments.
Cruising World magazine, USA, Apr. 1980 – Article by Mike O'Brien.
WoodenBoat magazine, USA, Jan./Feb. 1988 – Article by Mike O'Brien.
WoodenBoat magazine, USA, Plan No. 400-093 – Full building plans and offsets.

SOLWAY (DUNLIN)
YM Nov. 1908 – Lines, Sail Plan, Descriptive letter by Strange.
YM Feb., Mar., Apr. 1912 – Comments by first owner, Strange, and others on centreboards.
YM Mar. 1918 – Article by first owner, mentions longer bowsprit and increased sail area.
Mystic 1.691 – Lines, Sail Plan.

QUEST II
YM Jan. 1909 – Photographs and letter from first owner.
YM Nov. 1910 – Description of *Quest II* by editor.
YM July 1911 – Visit to Solway Firth by *Quest I.*
YM Apr. 1913 – Photograph and letter on canoe sterns by first owner.

CLOUD
YM Nov. 1908 – Lines, Accommodation, Sail Plan.
YW Feb. 1945 – Cruise Barcelona to Mallorca.
YW Nov. 1945 – Cruise Burnham on Crouch to Lisbon.
YW May 1946 – Cruise Lisbon to Valencia.
YW May 1947 – Return passage to UK.
Mystic 1.711 – YM Competition entry: Lines, Accommodation, Ketch Sail Plan.
             The design as built 1912: Lines, Table of Offests, Construction Plan,
              Yawl Sail Plan, Accommodation Plans, proposed and modified.

BLUE JAY
RCC Journal 1927 – Cruise Dartmouth to Orkney and return.
RCC Journal 1929 – Cruise around Ireland.
Eternal Wave. John Scott Hughes. London: Temple Press, 1951. Above cruises and others made by T N Dinwiddy.

BETTY. Later named **TALLY HO**
YM June 1910 – Lines, Construction/Arrangement, Sail Plan, and sections of the
Accommodation.
YM Oct. 1927 – Winning Fastnet Race when named *Tally Ho*. Also Lines etc. as above.
British Ocean Racing. Douglas Phillips-Birt. London; Adlard Coles, 1960 – Comments.
YM Oct. 1970 – Account of *Tally Ho* being stranded on the Hervey Islands and later towed to
Rarotonga to be repaired.
Mystic 1.695 – Lines, Construction, Mid Section, Accommodation, Sections of Accommodation,
Sail Plans, original and amended.

DESIRE
YM Jan. 1909 – Competition Rules.
YM Mar., Apr. 1909 – Lines of various entries including *Desire*.

PUFFIN II
YM Apr. 1910 – Lines, Construction/Arrangement, Sail Plan, Details.
Mystic 1.719 – Lines, Construction, General Arrangement, Sail Plan.

NORMA/SHULAH
YM Jan., Apr. 1936 – Wishbone Rig.
YM Jan., Feb., Apr. 1941 – Loss in gale.
Fore and Aft Craft. E Keble Chatterton. London; Seeley, Service and Co., 1922 and *Good Boats*.
Roger C Taylor. USA; International Marine Publishing Co., 1977. Both contain Lines,
Construction/Arrangement and Sail Plan.
Mystic 1.750 – Lines, Construction/Arrangement, Sail Plan.

IMOGEN II
The Complete Yachtsman. B Heckstall-Smith and E Du Boulay. London; Methuen, 1912.
USA; Dutton, 1921; various later editions – Lines, Construction/Arrangement, Sail Plan.
YM Oct. 1930 – 'Voyage to Las Palmas'. Details of running sails.
Single Handed Sailing. Richard Henderson. USA; Highmark Publishing Ltd., 1988.
London; Adlard Coles, 1989.
Wind Aloft, Wind Alow. Marin-Marie. London, Peter Davies, 1945. Both comment on
Captain Waller's Running Sails.
Mystic 1.687 – Lines, Construction/Arrangement, Sail Plan.

CHERUB III Canoe Stern Version
HYC Year Book 1911 – Lines, Construction/Arrangement, Sail Plan, Designer's comments.
The Corinthian Yachtsman's Handbook, 1913 and *Cruising Hints* 1928 and 1935 Editions.
Francis B Cooke. London; Edward Arnold. All contain folding drawings of Lines and Sail Plan.
Fore and Aft Craft and *Good Boats* (see *Norma* above) and also *Traditions and Memories of American
Yachting*. W P Stephens. USA; International Marine Publishing Co., 1981 and WoodenBoat Publications
Inc. 1989. JL page 64. All contain Lines, Construction/Arrangement, Sail Plan and comments.
Mystic 1.730 (Canoe Stern) – Lines, Construction/Arrangement, Sail Plan.

CHERUB III Transom Stern version
Yacht Cruising. Claud Worth. London; J D Potter. 1910. Lines, Construction/Arrangement, Sail Plan.
Mystic 1.730 (Transom) – Lines, Construction/Arrangement, Sail Plan.

ARIEL

YM Feb., Mar. 1929 – Cruise to Spain and return.

YM May 1929 – Article in 'The Other Man's Boat' series with Lines and Accommodation.

RCC Journal 1928 – Award of Romola Cup for cruise to Spain.

Little Ship Wanderings. J B Kirkpatrick. London; Edward Arnold & Co. 1933. Various cruises.

YM Feb., Mar. 1953 – Cruise to Holland and return.

GALATEA

YM Dec. 1937 – A cruise in the lowlands (Holland).

GRETTA

YM Dec. 1954 – Report of loss.

FIREFLY

YM Nov., Dec. 1922 – Specification. Lines, Construction/Arrangement, Sail Plan, as amended by T Harrison Butler.

The Motor Boat magazine, Aug. 1922 – Details similar to YM above.

The Cruising Association Bulletin, Dec. 1938 – Cruise to Denmark.

No. 119

YM July 1911 and JL page 117f. – Lines, Construction/Arrangement, Sail Plan.

Mystic 1.744 – Lines, Construction/Arrangement, Sail Plans, Sloop and Yawl.

AFREET

YM Dec. 1911, Jan., Feb., 1912 – Lines and Sail Plans of competition entries.

Mystic 1.718 – Lines, Construction, Sail Plan.

YM Dec. 1913 – Lines, Construction, Sail Plan. RCYC one-design.

FLAPPER

YM Nov. 1912 – Competition details.

YM Jan. 1913 – Commended entries.

Mystic 1.747 – Lines, Construction.

MOTH II

YM Aug. 1923 – Cruise Plymouth to Conway. Specification, Lines, Construction/Arrangement, Sail Plan.

Mystic 1.721 – Lines, Construction/Arrangement, Sail Plan, Table of Offsets.

8½ TON YAWL (NIRVANA)

YM June 1918 – Specification, Lines, Construction/Arrangement, Sail Plan.

YM Apr. 1929 – Review of *Nirvana*, 'The Other Man's Boat' series.

YM Aug, Sept. 1933 – Cruise Dublin to Plymouth and return.

Mystic 1.724 – Lines, Construction, Accommodation, Sail Plan.

No. 114
Yacht Cruising. Claud Worth, published London; J D Potter. Four editions, 1910, 1921, 1926, 1934. Suggested Specification, Lines, Construction/Accommodation, Sail Plan.
Of Yachts and Men. William Atkin. New York; Sheridan House, 1949, Reissued 1984. UK; Ashford Press Publishing, 1985.
Saga of Direction. Charles H Vilas. New York; Seven Seas Press, 1978.
Both discuss developments of the Colin Archer type of double-ender including AS No. 114.
Mystic 1.709 – Lines, Construction/Arrangement, Sail Plan.

No. 147
YM May 1916 – Lines, Construction/Arragement, Sail Plan, Text by Strange.
YM July 1916 – Perspective drawing of hull.
Mystic 1.737 – Lines, Construction, Accommodation, Sail Plan.

TUI
YM Apr. 1917 – Lines, Construction/Arrangement, Sail Plan.
Mystic 1.726 – Lines, Construction/Arrangement, Sail Plan.
YM July 1912 – Lines, General Arrangement, Sail Plan of *Tarana*, designed by Norman Dallimore for Walter Hopkins.

TWO DESIGNS TO B.R.A. RULE
YM Dec. 1916 – Article by Strange with Lines Drawings of designs.

NURSEMAID AND BABY
YM June 1917 – Article by Strange with Lines, Sail Plans and Cabin Sections of both designs.

THE JILT
YM Dec. 1918 – Specification, Lines, Construction/Arrangement, Sail Plan.
Mystic 1.681 – Lines, Construction/Arrangement, Sail Plan, Table of Offsets.

Chapter Four – An Old-Time Cruiser

An Old-Time Cruiser - YM April 1916.
Two Summer Cruises with the Baltic Fleet, in 1854-5; being the Log of the Pet *Yacht, 8 Tons, RTYC.*
Robert Edgar Hughes. London; Smith, Elder, 1855. 2nd Ed, revised 1856.

Vanderdecken. Pseudonym of William Cooper.
The Yacht Sailor: A Treatise on Practical Yachtsmanship, Cruising and Racing. Various editions, London; Hunt & Co.
1st Ed. 1860. *Yarns for Green Hands.*
2nd Ed. 1862. *Being a Reissue of Yarns for Green Hands.*
3rd Ed. 1868 and 4th Ed. 1874. *The Yacht Sailor: A treatise on Practical Yachtsmanship, Cruising and Racing.*
5th Ed. 1876. 3rd and 4th Eds. with notes and additional chapters by a Clyde yachtsman.

Yachts and Yachting. Being a Treatise on Building, Sparring, Canvassing, Sailing and General Management of Yachts. With remarks on Storms, Tides. London; Hunt & Co. 1st and only Ed. 1873. An early example of the genre later exemplified by Claud Worth's *Yacht Cruising*, Francis B Cooke's *Cruising Hints* and Eric Hiscock's *Cruising Under Sail.*

Chapter Five - A Winter's Tale

A Winter's Tale – YM May, June 1914.
In 1910, Charles Pears, a notable Marine Artist and former pupil of Albert Strange, wrote and illustrated a book describing a cruise in his own small yacht over much the same ground as that covered in 'A Winter's Tale'. This was entitled '*From the Thames to the Seine*' and published in London by Chatto & Windus.

The Hills and the Sea. Hilaire Belloc. London: Methuen, 1906. A series of essays, five of which have some small boat content. The quotation mentioned by Strange is taken, incomplete, from the essay entitled 'The North Sea'.

On Sailing the Sea: A collection of the Seagoing Writings of Hilaire Belloc, made by W N Roughhead. London; Hart-Davis, 1951. Includes the five essays mentioned above.

Chapter Six - A Cruise on the Elbe and Baltic

A Cruise on the Elbe and Baltic – YM July, August 1908.
HYC Yearbook 1898 – Short account of same cruise, also by AS.
YM May 1908 – Account of an earlier cruise made by G Holmes in Danish Waters in 1894.
YM April 1909 – Another cruise to the Baltic made by G Holmes in 1908, again in *Eel*.
Both written and copiously illustrated with sketches by G Holmes (eg below).

The Falcon on the Baltic; a Coasting Voyage from Hammersmith to Copenhagen in a Three-Ton Yacht. Edward Frederick Knight. 1888. Later edition, 1915 London; Longmans. 1951, London; Rupert Hart-Davis, Mariners Library series with introduction by Arthur Ransome.

A Yacht Designer's Library 1851-1913

Although many books on all aspects of naval architecture were published during Albert Strange's lifetime, few were concerned solely with yacht design and fewer still with small cruising yachts.

Unfortunately, Strange's library, which had been given to the then Scarborough Sailing Club on his death, was lost when the premises were flooded during WWII. Thus, we will never know exactly which design books were among his collection, apart from Dixon Kemp's *Manual of Yacht and Boat Sailing* and W H White's *Manual of Naval Architecture,* to which he refers in the text of his design articles and elsewhere, and Folkard's *The Sailing Boat*, a copy of which was presented to him on his being elected the first Honorary Life Member of the Humber Yawl Club. He had certainly studied the subject in depth (see JL page 101) so it is likely that he would have been aware of the volumes listed below even if they did not all have a place on his book shelves.

Many of these books can still be found today through the catalogues of specialist dealers and in the reference departments of the National Maritime Museum, Greenwich; the Science Museum, South Kensington; and the Cruising Association Library at their new headquarters in Limehouse, London.

NOTES ON YACHTS. Edwin Brett. London; Sampson Low, Son & Marston, 1869.
Although described on the title page as 'First Series' there is no evidence that any further series was ever published. A concise, non-technical, discussion of existing yachts, hull shape, sails and rigging. A prophetic final chapter entitled 'Very Small Yachts'.

THE LAWS OF DYNAMICS APPLICABLE TO YACHT BUILDING. H C Chapman. London; C Wilson, 1851.

HOW TO DESIGN A YACHT. C G Davis. New York; The Rudder Publishing Co., 1906.
Charles G. Davis was a naval architect and writer whose work appeared in the American magazines *The Rudder* and *Motor Boat.* In 1905, or thereabouts, he designed a small canoe yawl whose lines were featured in Weston Farmer's book *From My Old Boat Shop*, International Marine Publishing Co., USA. 1979. It is interesting to compare this design with *Sheila* and *Mist*, drawn by Strange at about the same time.

THE SAILING BOAT. H C Folkard. Editions in 1853, 1854, 1863, 1870 and 1901. Re-issued in 1906. Various publishers.
The first four editions were in a small format and are very inferior to the massive fifth edition published by Edward Stanford, London, in 1901 and re-issued in 1906. A veritable gold mine of designs. Restricted Classes, one-designs, Raters, canoes, foreign and colonial boats. When Strange was elected as the first honorary member of the Humber Yawl Club in 1901 he was presented with a copy of this book. It contains two of his designs, *Otter* (JL page 48) and *Wenda*, see page 119f.

YACHT DESIGNING. Dixon Kemp. London; *The Field* Office, 1876.
A folio-sized volume with the sub-title 'A Treatise on the Scientific Principles upon which is Based the Art of Designing Yachts'. The text is illustrated with many folding drawings of vessels, large and small, including the schooners *Egeria* and *Sea Belle* on which Strange sailed as a boy (JL page 14).

YACHT ARCHITECTURE. Dixon Kemp. London; H. Cox, 1885.
This is a re-written version of *Yacht Designing*, above. Further revised editions appeared in 1891 and 1897.

A MANUAL OF YACHT AND BOAT SAILING. Dixon Kemp. London; *The Field* Office, 1878.
In addition to being an Associate of the Institute of Naval Architects, Dixon Kemp was also the Secretary of the Yacht Racing Association, and Yachting Editor of *The Field* magazine. He was a most influential figure in British yachting and this classic book ran to eleven editions. Each of these contained drawings of existing craft and it is known that at least one of Strange's early designs was based on a drawing from Dixon Kemp. In 1988 John Leather edited a completely revised version of the original eighth edition and copies are still available – published by Ashford Press Publishing.
The 11th edition, published in 1913 (Dixon Kemp died in 1899), was edited by Brooke Heckstall-Smith and Linton Hope MINA. This was in two volumes and combined the works *Yacht and Boat Sailing* and *Yacht Architecture*.

SMALL YACHTS – THEIR DESIGN AND CONSTRUCTION. C P Kunhardt. New York;
Forest and Stream, 1891 and London; Sampson Low, 1891.
Another folio-sized volume. Covers the principles of yacht design with many very clear drawings of both British and American examples. Sections on Single-Handed Sailing, and Seamanship.
An abridged edition was published in the USA by *WoodenBoat* Publications Inc.

YACHTS AND YACHT BUILDING. P R Marett. London; E & F N Spon, 1856. 2nd Edition 1872.
This is believed to be the first comprehensive book on the principles of yacht design.
An attempt to persuade owners and prospective owners of the value of building to plans rather than relying on builders' models. Contains various folding plans including a scale drawing of the schooner *America*. Marett, who was a barrister by profession, was possibly the first of the long line of amateurs who have influenced the development of yacht design.

A TREATISE ON SHIPBUILDING, NAVAL ARCHITECTURE, ETC. Lord Robert Montagu. 1852.
Not as detailed in its coverage of yacht design as Marett, above.

ELEMENTS OF YACHT DESIGN. Norman L Skene. Date of 1st Edition not known.
2nd Edition 1909. The Rudder Publishing Company, USA.
A much handier size than some of the tomes listed above. In his preface the author says: 'The book is intended to be a practical and concise presentation of some of the operations involved in designing yachts of all types. Cumbersome and impractical methods which are often found in more pretentious works on naval architecture have been avoided. Those presented have been in everyday use by the author'.
Norman Skene had worked in the design offices of both Starling Burgess and L Francis Herreshoff.
This book has become a standard work on the subject. The eighth edition (1973) was revised and updated by Francis S. Kinney.

A MANUAL OF NAVAL ARCHITECTURE, FOR THE USE OF OFFICERS OF THE ROYAL NAVY, OFFICERS OF THE MERCANTILE, MARINE SHIPBUILDERS, SHIPOWNERS AND YACHTSMEN.
W H White. London; John Murray, 1882.
C P Kunhardt, editor of *Small Yachts* (see above), described this book as 'The modern standard work of the kind, well fitted for study by yachtsmen'. At least four further editions were published.

DESIGN ARTICLES IN MAGAZINES

In his book, *Cruising Yachts – Design and Performance*★ the influential amateur designer, Harrison Butler, pays tribute to a series of articles by the Glasgow naval architect Mr J Paterson which commenced in *The Yachtsman* in January 1896.

A somewhat incomplete but, nevertheless, useful series was written by Herbert Reiach, the founding editor of *The Yachting Monthly*. These were published in YM Nov. and Dec. 1912, and Jan., Mar., May 1913.

★First published in 1945 and long out of print. A new, enlarged, fourth edition has now been produced. *Cruising Yachts – Design and Performance*. T Harrison Butler. London; Excellent Press, 1995.

APPENDIX 3

The Albert Strange Association

To determine the birth of the Albert Strange Association, we have to go back to 1975, when the Honorary Secretary of Scarborough Yacht Club decided to research the early years of the Club with a view to producing a history. In reading the Scarborough newspapers spanning the period 1885 to1976, one name, that of Albert Strange, stood out prominently in the early years and it soon became obvious that he was a remarkable man in many ways.

In December 1976, an article appeared in *The Yachting Monthly* headed 'Before the Deluge' in which Albert Strange's connection with the Humber Yawl Club was mentioned. Scarborough Yacht Club's secretary was moved to write to the editor to say that the S.Y.C. was proud to record that Albert Strange was, in fact, a founder member of the club, and had been chiefly responsible for its formation.

That letter opened the flood gates – a number of readers replied, and in the spring of 1978 sufficient progress had been made for a weekend meeting to be arranged in Scarborough at which the Albert Strange Association was formed. Unknown to these founder-members, there were many other enthusiasts in the UK and overseas who cherished and still regularly sailed designs from Strange's drawing board. The Association now has members all over the world. Not all own Strange designs or paintings, of course, but each is an enthusiastic admirer of his work.

This book, together with John Leather's earlier book, represent milestones in what is a continuing aim of the Association: 'To trace and preserve the designs and little ships of Albert Strange, and to make a permanent record of his life and work.' Many members have contributed to this aim, and invaluable co-operation has been received from Mystic Seaport Museum, Connecticut, which is a member of the Association and holds the W P Stephens Collection of Albert Strange plans. The support of readers of like mind is always welcomed - any who would like to join the Association are invited to write to:

Pete Clay
Honorary Secretary
Albert Strange Association
Saxon House
83 Ipswich Road
Woodbridge IP12 4BT
UK
01934 384374

This is not an owners' Association. Anyone who shares the members' interest in our heritage and its preservation can help the Association in its task. The Honorary Secretary is particularly keen to hear from anyone who knows of an Albert Strange design or boat which may not have been traced by the Association.

APPENDIX 4

CONVERTING IMPERIAL MEASUREMENTS TO METRIC UNITS

For the benefit of those readers who wish to convert from Imperial to Metric Units the necessary figures are given below:

Multiply feet by 12 and add any inches. Divide the total by 39.37. The answer is in metres.

1 hundredweight (cwt) = 112 pounds (lbs)   1 ton = 20 cwt = 2240 lbs

To convert lbs to kgs divide by 2.2

To convert cwts to kgs multiply by 50.8

To convert tons to metric tonnes multiply by 1.016

To convert square feet to square metres multiply by 0.0929

INDEX